SOOTHING THE ESTABLISHMENT

The Impact of Foreign-Born Scientists and Engineers on America

David S. North

Supported by the Alfred P. Sloan Foundation

University Press of America, Inc.
Lanham • New York • London

Copyright © 1995 by
University Press of America,® Inc.
4720 Boston Way
Lanham, Maryland 20706

3 Henrietta Street
London, WC2E 8LU England

All rights reserved
Printed in the United States of America
British Cataloging in Publication Information Available

Library of Congress Cataloging-in-Publication Data

North, David S.
Soothing the establishment : the impact of foreign-born scientists and
engineers on America / David S. North
p. cm.
Includes index.
1. Science--Study and teaching (Graduate)--United States. 2.
Engineering--Study and teaching (Graduate)--United States. 3.
Students, Foreign--United States. 4. Scientists--Supply and demand--
United States. 5. Engineers--Supply and demand--United States. I.
Title.
Q183.3.A1N67   1995   306.4'5'0973--dc20   95-3988 CIP

ISBN 0-8191-9887-0 (cloth.: alk. paper)

⊖™ The paper used in this publication meets the minimum
requirements of American National Standard for Information
Sciences—Permanence of Paper for Printed Library Materials, ANSI
Z39.48–1984.

Contents

Chapter One: An Overview 1
   A Specialized American Vacuum 5
   The Numbers of Foreign-Born Scientists and Engineers 8
   Characteristics of Foreign-Born Scientists and Engineers .. 12

Chapter Two: Motivations 21
   Americans' Graduate School Choices 21
   Graduate School Choices of Foreign Students 30
   The Different Levels of Interest in the Life Sciences 33

Chapter Three: The Gatekeepers 37
   The Educational Testing Service 37
   Graduate School Admissions 46
   Getting A Job in the U.S. 50
   The Immigration Process 55

Chapter Four: Roles of Foreign-Born
  Scientists and Engineers 69
   Overview 69
   The Foreign Born in Academe 71
   The Foreign Born in Industry: The Older, Larger Pattern . 101
   The Foreign Born in Industry: The Newer, More Controversial
      Pattern 111

Chapter Five: Impacts of Foreign-Born
   Scientists and Engineers 121
   The Impact on Education 122
   The Impact on Industry 129
   The Impact on U.S. Populations 133
   The Impact of the Foreign Born on
      Science and Engineering, *Per Se* 144

Chapter Six: U.S. Reactions to the Impacts of
   Foreign-Born Scientists and Engineers 145
   The Mainstream Reactions 145
   Eddies 152

Chapter Seven: Conclusions and Recommendations 161
   Conclusions 161
   An Alternative Vision 162
   Recommendations 164

Appendix 175

Index 179

Table of Exhibits

Exhibit One: Measures of the Stock and Flow of Foreign-Born
Scientists and Engineers 9

Exhibit Two: Country of Citizenship of Non-USC Recipients
of Science or Engineering Doctorates 16

Exhibit Three: Changing Financial Reward Patterns
Among U.S. Professions 24

Exhibit Four: Trends in U.S. Graduate
Degrees (1970 - 1990) 26

Exhibit Five: Differential Earnings Over Time
of Bachelor's Only and PhD Chemical Engineers--
Or Why Many Avoid Grad School 28

Exhibit Six: GRE General Test Scores for
Selected Populations: 1987-1988 41

Exhibit Seven: General GRE Mean Test Scores for
Would-be Graduate Students in
Science and Engineering, 1993 43

Exhibit Eight: 142 PhD Programs in Biochemistry, Economics, Mathematics, and Mechanical Engineering at 43 U.S. Institutions According to the Percent of Applicants and of Admitted Applicants who are U.S. Citizens 51

Exhibit Nine: Nonimmigrant Admissions of Scientists and Engineers, FY'93, by Visa Class 64

Exhibit Ten: Some Indirect Measures of Foreign-born Students' Prowess in Graduate School 77

Exhibit Eleven: Debt Status of New PhDs in Science and Engineering, 1991 85

Exhibit Twelve: How Postdocs' Weekly Wages Compare with Those of Male Workers Generally, 1990: BLS and Survey Data 95

Exhibit Thirteen: The Postdoc Labor Force, 1980-1991; Growing Rapidly With an Even More Rapid Increase Among the Foreign Born 97

Exhibit Fourteen: Civil Status of Assistant Professors of Engineering, 35 Years of Age or Younger, 1973-1989 99

Exhibit Fifteen: Naturalized Chemists Earn a Bit More than US-born Chemists, 1990-1992: Results of Salary Surveys Among ACS Members Holding PhDs 108

Exhibit Sixteen: Occupations Covered by the Department of Labor's Proposed LMI Program, An Easing of the Labor Certification Program 156

Chapter One

An Overview

Beginning in the 1970s and continuing into the mid-1990s, there has been a steady growth in the number and proportion of foreign-born scientists and engineers in America's graduate schools, on its university faculties, and in its corporations.

As an extreme example of this growth, we can look at the group receiving U.S. PhDs in civil engineering in 1991 — 71% of them were foreign born. The proportion of foreign born among those receiving PhDs in mechanical and electrical engineering was almost as high; similarly, in most recent years, a majority of those receiving U.S. PhDs in mathematics have been from overseas.

These trends are large enough and are in such a strategic segment of American life that much has been written about them. There have been studies of the academic part of this process — what does all this mean to U.S. higher education? There have been voices from the Third World concerned with the Brain Drain. And there are many U.S.-based descriptive and census-like studies, with more attention paid to *gathering* detailed information than to thinking about it. However, little has been written about the foreign-born scientists and engineers in the labor market.

The approach here is different — we are interested in the processes involved, the motivations of the players, the actions of the various gatekeepers, and primarily the direct and indirect impacts of this phenomenon on U.S. populations and institutions, including the labor market.[1]

The approach outlined above runs against the grain for many:

- Academics resist the concept that a science or engineering graduate school can be viewed as a labor market, and a nonregulated one where many highly-motivated and highly-educated people get rock-bottom salaries.

- Most scientists and engineers think of themselves as professionals, as special people who should not have to organize to advance their own economic interest.

- And everyone is uncomfortable with the possibility that an immigrant population may cause, in effect, the limitation of opportunities for U.S. women, Blacks, and Hispanics, or that industry may have an active preference for young male foreigners over middle-aged American citizens.

In this setting, it is useful to recall that employers, since the dawn of the industrial revolution, have often decided to change the labor force rather than to increase wages, or to change work practices, or to mechanize when they have difficulty finding the workers they want at the rates that they are willing to pay.

Sometimes these changes are helpful to population groups within the society, as employers drop old prejudices against employing women, Blacks, or the disabled. Sometimes these changes are helpful to populations (at least initially) outside the society, as aliens are admitted to do work previously unavailable to them.

Usually the work involved is difficult or dangerous or otherwise unattractive to residents (at the wages offered). One thinks of Poles in Belgium's coal mines before World War II, the current use of Algerians to sweep the streets of Paris, and the Jamaicans cutting sugar cane with machetes in Southern Florida; each of these groups of migrants replaced native-born work forces. Typically when employers succeed in changing the labor force as outlined above, they do so in order to retain the status quo in wages, working practices,

---

[1] The Brain Drain question is an important and interesting one, but is beyond the scope of this volume.

and/or lack of mechanization.

While the apparently unattractive work in the U.S. we are discussing is mental, not physical, and while the salaries paid for it are in the comfortable to generous range, do the same dynamics apply? Has the U.S. opted to import foreign workers in science and engineering to retain the status quo? To keep the grad schools busy and the industrial salaries at current levels? To avoid a different distribution of income that might adversely affect the ruling elites?

Before seeking to answer these questions, it is useful to outline here, and describe more thoroughly in the chapters that follow, the basic patterns operating in the field.

In the last forty years or so, the American labor market, presumably reflecting society's values, has increased financial rewards to physicians, lawyers, and business executives while decreasing the compensation (relatively) to scientists and engineers. This is happening in a society where making things has become less important and providing services and manipulating money and systems has become more important.

As a result of these trends, a smaller percentage of Americans who complete four years of college seeks graduate degrees in science and engineering and a larger percentage opts for degrees in medicine, law, or business, as we show subsequently. Further, an American engineer with a bachelor's degree is immediately employable at a relatively good salary and need not seek a graduate degree to be regarded as a professional. Graduate science and engineering departments, thus, need students and faculty (or face contraction) and industry still needs graduate scientists and engineers, leaving a potential partial vacuum that can be filled by the foreign born.

Although spending five or more years at an American university doing difficult work on a low stipend attracts only a limited number of qualified Americans, it has a much stronger appeal to equally (if not more) qualified foreign students. To them an American PhD is an outstanding credential, and the prospect of five years in an American university is that of a real adventure.

Thus, many able foreign students seek admission to U.S. graduate schools, large numbers are admitted, and most of those admitted secure the degrees sought. U.S. employers offer jobs to most of the foreign born with U.S. (or Canadian) degrees, and most of the foreign-born graduate students decide to stay in the U.S. The U.S. immigration system, while sometimes the source of delay or other irritations, generally facilitates these processes, a point to which we will return later.

A very distinct path is used by most overseas scientists and engineers (S/Es hereafter) as they come to the U.S. We call it the immigrant path. Typically they do not secure their bachelor's degrees in North America; this is done in the homeland. If they do well at university, they then apply to U.S. or Canadian graduate schools, for either a PhD or a master's. (The latter, while it may be useful elsewhere in the world, is more often regarded as an excellent stepping stone to a North American job.) So the typical foreign-born scientist or engineer arriving in the North American labor market has a brand-new North American graduate degree and at least a couple of years' experience with local systems and expectations. There has been some migration of mature and often stellar scientists in recent years — such as the Russian mathematicians and physicists — but these movements are one-shot affairs and involve much smaller populations than the ongoing pattern described above.

There is another basic pattern, the clearly exploitive employment of run-of-the-mill alien engineers and computer programmers who arrive in the U.S. in groups, often illegally. This, which we call the nonimmigrant path, is described separately.

The rest of this chapter sketches the numbers and characteristics of scientists and engineers in the U.S., particularly the foreign-born. The following chapters deal with the motivations of the various players, outline the activities of the gatekeepers they encounter, describe the role of the foreign-born S/Es in academe and in industry, and discuss the direct and indirect effects of this group on U.S. institutions, notably universities and corporations, and on U.S. populations (including native-born women, Blacks, and Hispanics). We then offer some observations about the reactions within the U.S. to these trends, and we close with some recommendations.

A definitional note is needed regarding citizen/alien status. There are four basic groups of interest:

- Native-born U.S. citizens (USCs)
- Naturalized USCs
- Immigrants or permanent resident aliens (PRAs)
- Nonimmigrants, or aliens, in the U.S. on temporary visas

Nonimmigrants, either by marrying a USC or PRA, or by

securing a permit to work permanently in the U.S. (a labor certification), may apply for immigrant status. After three to five years (depending on marital status), an immigrant may apply for naturalization.

Unfortunately, data sets differ. Some distinguish among all four classes; some lump citizens and noncitizens into two groupings. Others make a distinction between those with temporary visas and those who are either citizens or resident aliens (Canadian data sources regularly use the last option). We use foreign born to cover naturalized citizens, immigrants, and Nonimmigrants.

Fortunately, when dealing with the immigrant path, two of the four classes (naturalized citizens and PRAs) are *not* numerous during graduate school, when the foreign born are largely in nonimmigrant status. Similarly two of the four groupings are *not* very significant in the labor market phase of the immigrant pattern because most of the foreign born who stay in the U.S. seek naturalization. Thus, there are more naturalized citizens among the foreign-born S/Es than there are PRAs and temporary visa holders combined. All of those in the nonimmigrant path are on temporary visas and have little or no opportunity to change that status.

A Specialized American Vacuum

One reason why there is an increasing number of foreign-born scientists and engineers at the PhD level is because a smaller proportion of Americans have sought science and engineering PhDs in recent years than in the past.

If one looks at science and engineering employment in the U.S. with a telescope one sees a large and growing professional work force, with only a modest contribution of foreign workers. For example, according to the U.S. Census, the total number of people working in science and engineering occupations rose from 1,936,000 in 1980 to 3,126,694 in 1990, with the foreign-born percentage being a not overwhelming 11.8% in 1990. One might compare this to the foreign-born percentage in the population as a whole (8.7%) in 1990, and detect a noticeable but not remarkable difference.

The 1990 Census data show the following about the three major subclassifications in the field:

Profession	Total Population	% Foreign Born
Engineering	1,904,565	12.3%
Math./Computer Science	802,329	10.0%
Natural Science	419,800	12.6%
Total	3,126,694	11.8%[2]

But if one uses a microscope instead of a telescope to look at the best-educated minority among this professional work force, one sees a different picture, as native-born Americans move away from the hard work of securing a science or engineering PhD and as the foreign born move in to fill this vacuum. It is at the top of the science and engineering pyramid where the foreign born are most likely to be found.

With this in mind, let us examine two birth cohorts, Americans born in 1945, just before the post-World War II baby boom, and those born in 1958, at the height of the baby boom. Assume that a person is about 30 years old when he or she completes a PhD; let us now look at the number of science and engineering PhDs awarded to USCs in 1975 and 1988:

---

[2] The author is grateful to Dr. Leon Bouvier for extracting these data from the 1990 Census. For a comprehensive review of what the 1990 Census tells us about foreign-born professionals generally and the foreign born in science and engineering more precisely, see Leon F. Bouvier and David Simcox, *Foreign Born Professionals in the United States*, Center for Immigration Studies, Washington, D.C., 1994.

Discipline	1975	1988
Physics	925	721
Chemistry	1,395	1,373
Env. Science	492	507
Mathematics	848	341
Computer Science	0	284
Life Sciences	3,473	3,658
Engineering	1,716	1,778
Total[3]	8,846	8,662

Over a period of 13 years, the absolute totals for USCs dropped by about 2%. Compared to the total population of the country in the two degree-granting years, however, the proportion of new science and engineering PhDs dropped more sharply, from 41 for every million in the population in 1975 to 35 per million in 1988. But the decline was most notable when compared to the two birth cohorts. There were 31 of those new PhDs in 1975 for every 10,000 born in 1945, but only 21 per 10,000 thirteen years later, a decrease of about one third.[4]

Yet another measure of this declining American interest in science and engineering at the graduate level is the ratio between the number of USC PhDs in a given year and the number of bachelor's degrees (in all disciplines) awarded six years earlier. For every 10,000 graduates of American colleges in 1969, 115 USCs earned science or engineering PhDs six years later; for every 10,000 graduates in 1982, there were 91 science or engineering PhDs for USCs in 1988.[5]

---

[3] These totals, as elsewhere in this volume unless otherwise noted, do not include psychologists and social scientists.

[4] Doctorate data from National Science Board, *Science and Engineering Indicators— 1989*, Washington, D.C., U.S. Govt. Printing Office, p. 226; birth and population data from the *Statistical Abstract of the U.S.*, various volumes.

[5] Bachelor's degree data from *Statistical Abstract of the U.S.*, table 212 of the 1972 edition for 1969 data, and table 266 of the 1985 edition for 1982; the source of doctorate data was shown in the previous footnote.

But native-born Americans have not been shying away from science and engineering generally, just graduate school. Comparing birth cohorts to USCs receiving undergraduate degrees in 1977 and 1990, we find that of the Americans born in 1955, 40 in 1000 received science or engineering bachelor's degrees 22 years later while the ratio had moved up to 45/1000 for those born in 1968 and leaving college in 1990. The absolute numbers of the bachelor's degrees had dropped a little between the two years, from 162,548 to 160,694, but the birth cohort in 1955 was half a million larger than the one in 1968.[6]

The Numbers of Foreign-Born Scientists and Engineers

Two broad statements can be made about the population of foreign-born scientists and engineers in the U.S. First, the numbers keep growing no matter how you count them (several different systems are available), and second, the educational-attainment patterns of the foreign born are the mirror-image of the USC patterns described above. In 1990, for example, only 5.4% of the bachelor's degrees in science and engineering went to those with temporary visas, while 27.0% of the master's degrees were awarded to this population.[7] Of the PhDs awarded that year to people of known citizenship, 32.4% went to nonimmigrants.

As Exhibit One indicates, virtually all of the trends regarding foreign-born S/Es in the U.S. are going in a single direction — up. The exhibit shows two different flow measurements: Immigration and Naturalization Service (INS) data on immigrant and nonimmigrant admissions to the U.S. and counts of foreign born receiving three different academic degrees. It also shows measures of the stock of foreign-born S/Es in universities, generally, and in graduate schools, specifically.

Several additional comments can be made about the data in Exhibit One. First, there is the explosion in the employment of

---

[6] Data on degrees from National Science Foundation, *Foreign Participation in U.S. Academic Science and Engineering*, Washington, D.C., 1993, pp. 76-77; birth data from *Statistical Abstracts of the U.S.*

[7] NSF, *Foreign Participation, op.cit.*, pp. 76-77.

Exhibit One: Measures of the Stock and Flow of Foreign-Born Scientists and Engineers

Measure Used	1969	1979	1985	1990	1991	1992	1993
INS Data							
Nonimmigrant Admissions (FYs)	3,947	3,660	*28,106	39,583	52,992	54,653	59,664
Immigrant Admissions (FYs)	9,259	8,113	*10,746	13,572	13,742	21,782	22,300
Enrollment Data							
Graduate School	na	na	50,938	70,245	73,843	75,059	na
All University	51,137	114,220	136,960	142,540	145,740	147,850	153,750
Degrees Granted							
Bachelor's	na	7,566	11,201	9,243	9,259	na	na
Master's	na	6,792	10,266	13,569	14,531	na	na
Doctorates	2,450	3,149	4,335	6,957	7,408	8,204	na
Total	na	17,507	25,802	29,769	31,198	na	na

* FY '86

Note: INS and degrees-granted data measure the flow of foreign-born S/Es through INS and academic processes; enrollment data are a stock measure of nonimmigrants studying science and engineering in U.S. institutions of higher learning. Nonimmigrant admissions after 1979 and some enrollment information are estimates; the rest of data are counts. For an explanation of data sources and definitions, see the Appendix.

nonimmigrants generally over the last quarter of a century. Admissions in all occupations in the classes of interest (those that allow the visa holders to work legally) went from 96,000 in 1969 to more than 263,000 in 1990, about a three-fold increase.[8] But in this period the number of S/Es among these nonimmigrant classes increased ten-fold. Second, while the admissions of foreign-born S/Es as immigrants has increased, the trends are not as dramatic as those for nonimmigrants.

Third, the various enrollment and degree-granting data show a growing proportion of the foreign born seeking or attaining advanced degrees and a further lessening of interest in the bachelor's degree, which, as we have stated before, is usually earned by the foreign born before coming to the U.S.

Finally, it should be noted that there is a great deal of overlap among these data. A single nonimmigrant, enrolling for a bachelor's degree and subsequently getting a master's and a doctorate — and a green card — would be recorded at some point on the exhibit in all seven rows; that would be an exceptional pattern, however. But someone arriving with a working nonimmigrant visa (such as a J-1) to enroll in graduate school and subsequently securing both a PhD (or a master's) and a green card would be a quite typical pattern; this person would be recorded in five of the seven rows.

Presumably the growing stock in graduate school will be reflected later in growing numbers of doctorates and green cards awarded and a growing stock in the work force.

The relative stabilization in the number of native-born PhDs awarded in science and engineering in the last twenty years and the growing number of doctorates granted to the foreign born would suggest that the percentage of such degrees in the hands of the foreign born has increased, and it has. If we total newly-granted doctorates in the life, physical, and mathematical sciences and in engineering over the last thirty-some years, we find the following:

---

[8] These totals cover classes H-1 (temporary workers of distinguished merit and ability), H-2B (other temporary workers), H-3 (industrial trainees), and J-1 (exchange visitors) in 1969, and those classes plus L-1 (employees of multi-nationals) in 1990. Farm workers are excluded in both years. The class of foreign-born S/Es most likely to be exploited, those working on business visas (B-1s), are *not* included in Exhibit One. For the sources of these data see tables 16B (1969) and 41 (1990) in the INS annual reports.

	1960	1970	1980	1990
USC	83.2%	80.0%	74.2%	60.0%
Non-USC	16.8%	20.0%	25.8%	40.0%
Sci. & Eng.				
PhDs	4,675	13,755	12,051	17,364[9]

We have previously noted the relatively minor proportion of U.S. bachelor's degrees in science and engineering that go to the foreign born and the middling number of master's degrees. The foreign born not only concentrate at the graduate level, they also focus on certain disciplines, ones in which there is a declining or stable number of the native born. These were the total 1991 doctorates awarded in the three major groupings:

Discipline	**Non-USC**	**USC**
Engineering	2,854	1,977
Physical Science,		
incl. Mathematics	2,532	3,450
Life Sciences	2,022	4,629[10]

Interestingly, in the fields where the foreign born are the least well represented, Life Sciences, there is a much higher representation of not only native-born Americans, generally, but also of women, Blacks, and Hispanics, a subject we cover in Chapter 2.

In addition to the major flow of foreign-born S/Es described above — those completing their education in North America — there is a smaller flow of foreign-born S/Es who have concluded their education *before* migrating to the U.S. Because this population has

---

[9] These data are taken from the National Research Council's highly useful annual survey of newly-awarded U.S. doctorates, in this instance: Ries, P. and D. H. Thurgood, *Summary Report 1990: Doctorate Recipients from United States Universities*, National Academy Press, Washington, D.C., 1993, p. 8. (The report gives the results of data collected in the *Survey of Earned Doctorates* sponsored by five federal agencies: NSF, NIH, NEH, and the U.S. Departments of Education and Agriculture, and conducted by NRC.)

[10] *ibid* p. 13.

had no contact with the data-hungry U.S. academic establishment, much less is known about it.

The best estimate of the relative size of these two populations is from Michael Finn's work with the 1982 post-census survey of scientists and engineers funded by the National Science Foundation (NSF). Finn calculated that in 1981, 17,092 foreign national S/Es arrived in the U.S. labor market (some from overseas, some securing degrees in the U.S.; some PRAs, some nonimmigrants). Of this population, 3,328 or 19.5% had completed their education overseas, while the balance, 13,764 or 80.5%, had U.S. degrees.[11] These proportions — particularly given the growing nonimmigrant pattern of science and engineering labor force utilization — probably have changed in recent years.

Characteristics of Foreign-Born Scientists and Engineers

Let us turn from numbers for a moment. It is useful here to offer a highly impressionistic group picture of the foreign born holding (or seeking) science or engineering doctorates.[12] (Less is known about the foreign born with lesser American degrees, and even less about those who come directly to the U.S. labor market with foreign degrees.)

As a group, the foreign-born PhDs are generally male, well-prepared, hard-working, amenable to authority, largely Asian, and highly likely to become naturalized citizens.

All have survived a remarkable series of screening devices on their way to the U.S. and many come from elite families in the Third World. Their principal weakness (in this setting) is their somewhat shaky command of the English language.

Since Third World cultures are less supportive than ours about encouraging women to participate in science and engineering, it should come as no surprise that the percentage of women among the foreign-born doctorates and graduate students is below that of USCs.

---

[11] Michael Finn, *Foreign National Scientists and Engineers in the U.S. Labor Force, 1972-1982*, Washington, D.C., USGPO, 1986, p,37.

[12] This account is based on numerous interviews with academics and other observers, many interviews with foreign-born S/Es themselves, and some information that can be gleaned from the literature.

For example, while 13.5% of the engineering doctorates for USCs in 1991 went to women, only 5.5% of those going to nonimmigrants went to women. For all doctorates in the sciences (including, in this instance, psychology and social sciences), 39.5% of those for USCs were for women, compared to only 21.5% of those for nonimmigrants.[13] Many of the foreign-born women pursuing graduate degrees in science or engineering are said to have parents working in the field.

The academic preparation of the foreign born admitted to U.S. graduate schools in science and engineering is often said to be equal to or better than that of U.S. applicants. Many come to the U.S. after securing master's degrees in their home countries; others have been able to concentrate more narrowly in their major fields than is possible in U.S. institutions, which often have distribution requirements.

A citizen of the UK, completing his PhD in mathematics at MIT, told me that he had been able to study virtually nothing but math for the last two years prior to attending Cambridge University and then to continue these studies throughout his stay there. "None of my American colleagues had this much concentrated math," he said.

While most Third World educational institutions cannot compete with American and European ones regarding modern laboratory equipment, this lack is frequently regarded (by American academics) as a blessing in disguise. "You do not need machinery to teach the principles of science — just a room, a blackboard, and the books; and lacking equipment, the students secure a splendid grounding in theory." We heard several variations on this statement.

More specific were the stories about Russian physicists and computer specialists; because of the primitive nature of Soviet-era computers, people working with them had been forced to new heights of software ingenuity to cope with the inadequate hardware.

Foreign-born PhD candidates are universally regarded as very diligent, very dedicated to their work. "They do not have the distractions that affect Americans," I was told many times. "But then our people would be just as hard-working if they were seeking degrees in Moscow," one observer added. The cliché in academic circles is this question: "If you needed something done in the lab

---

[13] *Foreign Participation, op. cit.*, p.84.

during the Super Bowl, who would do it for you, an American or a foreign student?"

It was usually later in the conversation that the question of reaction to authority came into play. Sometimes it was a statement that most of the foreign-born graduate students worked well in hierarchial situations (like American universities) because they had been raised in hierarchial societies. More often it was a comment that the foreign born, as a group, were amenable to authority, an important quality in a labor market and a rarely-discussed subject, to which we will return.

The foreign-born graduate students who reach the U.S. have typically been screened through a lifetime, long-odds, survival-of-the-fittest educational system. Third World educational systems do not try to bring the entire population through high school and do not facilitate massive attendance in higher education. These hard-pressed systems are more likely to provide good education to the few than equal educational opportunity to the many. So it is the very best students, taken from the top ranks elsewhere in the world, who seek PhDs in the U.S.

An officer of one of the mathematical societies seemed, at first, to be using statistical hyperbole as he made this point. "You are dealing here with the top one thousandth of one percent of the Chinese population," he told the author.

Later we did some rough calculations. There were about 1,150,000,000 people in China in 1989. One thousandth of one percent equals 11,500. If one assumes that one-sixtieth of that population turned 30 that year, there would be 191 Chinese of PhD-receiving age in that select group in 1989. And that year, according to the National Research Council (NRC) there were 318 recipients of U.S. doctorates in physical science and mathematics from Mainland China.[14]

The screening process that these 318 survived prior to admission to U.S. graduate schools was not just an intellectual one. The parents were presumably located, if not in the top tiny fraction of the society, at least near enough to the top to permit their children to attend college, as opposed to working in the fields. Presumably, many of the foreign-born graduate students are, like the Chinese

---

[14] *Summary Report 1989, op. cit.*, p. 42.

PhDs, members of their homeland's elite.

The one handicap afflicting many foreign-born graduate students — the Chinese, for example, much more than the Indians — is a lack of proficiency with the English language. This is more of a problem in some disciplines than others (more in biology than in mathematics) and more of a problem for foreign-born graduate students (and their students) who are teaching assistants (TAs) rather than research assistants (RAs). (Testing for English proficiency is one of the challenges faced by the gatekeepers, a subject we cover in Chapter 3.)

A large majority (71%) of foreign-born PhDs in science and engineering are from Asia. A distribution by continents for 1991 shows the following:

Continent	Number of Doctorates[15]
Asia	6,018
Europe	929
North America	496
Africa	473
South America	372
Australia - N.Z.	54
unknown	162
Total	8,504

As Exhibit Two shows, there has been a dramatic increase in the number of Chinese securing science and engineering PhDs in the U.S. in recent years. While the number of doctorates for Taiwanese has kept increasing, there has been an explosion in the number of degrees awarded to people from the PRC. If one adds in the degrees awarded to persons from Hong Kong and those unwilling to be specific about which China should be assigned to them, there was a

---

[15] *Foreign Participation, op. cit.*, p. 85. The dominance of Asians shown above is blurred by the way this NSF publication arrays the nations of the world. Asia is divided into "East Asia" and "West Asia"; the Philippines and Indonesia are included in "Australasia."

Exhibit Two: Country of Citizenship of Non-USC Recipients of Science or Engineering Doctorates
(includes those in Psychology and Social Sciences)

Nation	1983	1987	1991	diff. '83-'91
China				
PRC	13	290	1,596	+1,583
ROC	604	787	1,082	+478
Hong Kong	98	87	115	+17
unspecified	0	9	141	+141
sub total	715	1,173	2,934	+2,219
South Korea	236	547	1,067	+831
India	383	503	719	+336
Canada	179	181	276	+97
Iran	278	182	211	-67
Greece	66	123	158	+92
Mexico	96	86	126	+30
U.K.	87	94	125	+38
Japan	103	83	115	+12
Germany	43	74	114	+71
Other nations	2,112	2,489	2,659	+547
Total: All nations other than the US	4,298	5,535	8,504	+4,206

Note: nations are arrayed by number of doctorates awarded in 1991.

Source: *Foreign Participation, op cit*, pp. 86-87. "Unspecified" are those who responded "China" when asked about their citizenship. Also, we use U.K. here although the original uses the term England.

grand total of nearly 3,000 PhDs awarded to Chinese from various jurisdictions in 1991. (Singapore, another Chinese-populated nation, apparently does not encourage its scholars to seek science and engineering degrees in the U.S.; it never appears on lists of the top 25 source countries.) In 1991 more than 35% of the non-USC science and engineering doctorates went to the Chinese as we have identified them above.

Korea, India, and Canada are the next three nations on the list, followed by Iran, the only country whose production of U.S. PhDs has diminished in recent years. The regime of the Shah was apparently more interested in securing a Western education for its students than the subsequent Islamic fundamentalist government.

Surprisingly, Greece is the ranking European nation, followed by the U.K. and Germany. Germans and Brits largely come to the U.S. to secure degrees in science, not engineering; for example, of the 114 degrees awarded to Germans in 1991, 107 were in the sciences and 7 in engineering.

The ratios of new U.S. PhDs to total population in the leading sending countries is interesting; the Taiwanese have a U.S.-degree-to-population ratio only slightly higher than the U.S. does. The much higher ratios for PRC and India show the extent of the talent pool from which the U.S. PhDs from those nations are selected. The ratio in India is 67 times as high as it is in the States.

Citizenship of new PhDs	Population (millions)	1991 PhDs (Science & Engineering)	Ratio[16]
USA	249	13,923	1/17,884
Taiwan	21	1,082	1/19,408
S. Korea	43	1,067	1/40,299
PRC	1,151	1,596	1/721,177
India	866	719	1/1,204,450

Over their lifetimes in the U.S., a little less than half of all

---

[16] Data on doctorates (including those in psychology and the social sciences) from *Foreign Participation, op. cit.*, pp. 84, 86, and 87; population data from the *World Almanac 1993*.

immigrants become naturalized citizens. Generally speaking those migrants who feel that they can not safely visit their home countries (e.g., Cubans and, for years, Vietnamese) are more likely to naturalize than those not estranged from their homelands (e.g., Canadians and Mexicans). Also, those with more education and those with higher incomes are more likely to naturalize than those with less education and those with lower incomes.[17] The foreign-born S/Es are among the most educated people in the U.S. Therefore, it's no surprise that there is a strong tendency for the foreign-born S/Es — or at least among those staying in the U.S. — to become U.S. citizens.

Michael Finn dug the following statistics out of NSF data; they deal with the extent of naturalization of foreign-born S/Es identified in the 1980 Census who were still in the U.S. at the time of the 1982 post-census survey:

Population	% Naturalized in 1982[18]
Foreign-born S/Es identified in the 1980 Census, still in the U.S. in 1982, whose date of entry was:	
1970-1974	53.5%
1960-1964	81.6%
1950's	92.9%
before 1950	99.9%
all years	70.9%

"Date of entry" is an INS term-of-art meaning the start of PRA (immigrant) status; one could arrive in the U.S. in nonimmigrant status in, say, 1958 and not become an immigrant until 1963. For others, the date of physical arrival in the U.S. and the date of entry are simultaneous.

---

[17] D.S. North, *The Long Gray Welcome: A Study of the American Naturalization Program*, NALEO Education Fund, Washington, D.C., 1985.

[18] Michael G. Finn, "Foreign National Scientists and Engineers in the U.S. Labor Force, 1972-1982" Labor and Policy Studies Program, Oak Ridge Associated Universities, Oak Ridge, TN., 1985, p.24

Since most immigrants wait seven or eight years after the date of entry before beginning the non-instantaneous naturalization process, it will be decades before we know if the current group of arriving foreign-born S/Es are as adamant about U.S. citizenship as those covered by the 1980 Census. What we do know is that INS statistics show a steady stream of naturalizations for foreign-born S/Es in recent years:

Fiscal Year	Foreign-Born S/E Naturalizations[19]
1982	6,590
1984	7,641
1986	9,946
1988	6,048
1990	6,479
1992	6,280

When naturalizations for the populations shown above are compared to those foreign-born S/Es with a date of entry eight years earlier (i.e., in the even years 1974 through 1984), we find a citizenship-acquisition rate of about 85%; when all naturalizations in the listed years are compared with all admissions eight years earlier the rate is 46.2%, indicating that the foreign-born S/Es are much more likely to naturalize than immigrants in general.[20]

We next explore the motivational systems that encourage talented

---

[19] Unpublished data from the INS Statistics Division; the categories covered in recent years are "engineers, surveyors and mapping scientists," "mathematical and computer scientists," and "natural scientists"; for historical naturalization data by major occupation groups, see Table 49 in the *1991 INS Statistical Yearbook* and predecessor tables. We used every-other-year data here to save space. The surge in S/E naturalizations in 1986 related to a heightened INS effort that year to resolve a backlog of cases; naturalizations (more than 280,000 nationwide) reached an unusual peak at the time and apparently affected the S/Es as well.

[20] The percentages for 1980-1990 were calculated from *ibid*, Tables 1, 19, and 41, and for 1992, from as yet-unpublished INS data. S/E admissions data for 1979 were substituted for those of 1980, as INS lost much of it detailed immigration data that year; a further complication: INS occupational categories varied to some extent over these years.

foreigners to seek science and engineering PhDs in the U.S. and that tend to discourage Americans from doing the same.

Chapter Two

Motivations

In our study, three sets of motivations are of interest:

1. Why are Americans losing interest in graduate work in science and engineering in favor of LLBs, MDs, and MBAs?

2. Why do well-qualified foreigners flock to science and engineering graduate schools in the U.S. while shunning the LLBs, MDs, and MBAs?

3. Why are the life sciences more popular with Americans than the physical sciences and engineering and less popular with the foreign born?

Americans' Graduate School Choices

After four years of college most Americans go directly into the labor market; a minority go on to graduate school. Most decisions about graduate school relate to career choices — the new college graduate senses that additional specialized training and the degree attached to that training are needed to work in the field of choice.[21]

For some the motivation for an advanced degree is love of the field; for many it is a combination of interest in the subject and a

---

[21] In some fields the bachelor's degree provides professional credentials, as it does in engineering and pharmacy; in others, graduate work is needed.

desire to earn more than one could with a bachelor's degree alone. Four major factors go into many graduate school decisions:

- How many years will it take?

- How much will it cost, and who will pay for it?

- What will the value be, in increased earnings, of the advanced degree?

- What additional avenues of employment will it open?

A thumbnail answer to first question would be:

1 year:	Master of Library Science
2 years:	most other master's programs, including the popular MBA (Master of Business Administration) and MSW (Master of Social Work)
3 years:	LLB and JD, and most theological degrees
4 years:	DDS (dentists) and DVM (veterinarians)
6-7 years:	PhD in science or engineering (median registered-time-to-degree)[22]
7 years +:	MD (including post-MD internships)

Regarding the second question, graduate schools are expensive; with capital-intensive programs (such as those in science and engineering) being more costly to the university than others. The important variable to the potential student, however, is not the total cost of the program but the cost likely to be borne by the student. Again, in very general terms, financial support is readily available to most science and engineering PhD candidates,[23] available to some

---

[22] There are two measures used in the field: registered-time-to-degree is the length of time that the candidate is formally enrolled in the graduate school prior to receiving the degree; another (longer) measure is total time to degree, the number of years between the receipt of the bachelor's degree and receipt of the PhD. For more on this see *Summary Report 1991*, pp 14-15.

[23] For more on funding available to foreign-born and native- born science and engineering doctoral candidates, see Chapter 4.

MD and DVM students, and to few MBA or LLB students. In a rough way, the U.S. educational system provides more financial support for some graduate programs (such as science and engineering) where the ultimate salaries are relatively low and less financial support for those graduate programs that lead to higher income (law and medicine.)

The answer to the third question (the value, in increased earnings, of the degree) is that income differentials among occupations requiring post-college study have increased steadily in recent years. As Exhibit Three shows, in 1946 engineers, lawyers, and dentists had roughly the same incomes, though physicians made about 50% more than anyone else. By 1991 lawyers earned at least 50% more than engineers and scientists; the differential enjoyed by the medical profession was so dramatic that the Census Bureau apparently felt it necessary to blur the data with a statistical fig leaf (see note to the exhibit). An examination of earnings trends for 1983 and 1991 shows a continuation of the same differentials. These were the percentage increases, in current dollars:

Occupation	Percent Increase 1983 - 1991[24]
Physicians	98.0%
Lawyers and Judges	58.6%
Mathematical and Computer Scientists	44.8%
All Managerial and Professional Specialty Occupations	43.5%
All Engineers	41.2%
All Workers	37.4%
Natural Scientists	31.1%

The fourth factor is not a financial one; if one desires a career in teaching at the university level, one must secure a PhD. This is an attractive life for many, and the decision to seek a PhD in science or engineering is often influenced by this variable.

---

[24] Unpublished data from the Current Population Survey, provided by the Bureau of Labor Statistics over the telephone. The data for physicians covers physicians in salaried positions, excluding all self-employed; the excluded class probably has higher earnings than the covered class.

Exhibit Three: Changing Financial Reward Patterns Among U.S. Professions

A 1946-1991 Comparison of Median Earnings in Five Fields

(Basis: Median Engineers' Earnings = 100)

Year	Engineer	Lawyer	Physician	Dentist	Natural Scientist
1946	100	96	153	105	n.d.
early 1950's	100	109	191	160	n.d.
1959	100	128	176	144	93
1969	100	143	*190	165	91
1979	100	150	*204	n.d.	93
1991	100	154	*177	n.d.	91

* One of the intriguing elements that was encountered in preparing this table is the way that the Census Bureau, presumably under some pressure from the medical profession, has obscured the high earnings of physicians. In the 1970 Census (cited below) the only occupational category with an imprecise earnings figure was that of physicians, recorded as "$25,000 +"; it was also the highest. The 1980 Census data and the 1991 Current Population Statistics data no longer show earnings of physicians alone, instead their earnings are lumped with those of veterinarians, dentists and some other less-well paid groups, which tends to bring down the median. In other tables, however, Census carefully notes the exact numbers of physicians, dentists, etc.

Sources: 1946 and early 1950's data are from *Historical Statistics of the U.S.*, Series D 913-926 (data for engineers and lawyers are for 1953, for dentists, 1952, and for physicians, for 1951, which tends to understate the earnings differential for that group). Data for 1959 are from the *Statistical Abstract of the United States, 1966*, Table 328. Data for 1969 and 1979 are from 1970 Census of Population, *Occupation and Employment*, PC (2) 7-A, Table 1, and its companion volume for 1980. The 1991 data are from the Current Population Survey publication "Money Income of Households, Families and Persons in the U.S., 1991" and was provided over the phone by the Census Bureau. Comparisons are for males.

What happens when American college graduates look at the earnings, time-to-degree, graduate school cost, and career options data? The decision, for increasing numbers of them, is to seek degrees in fields other than science or engineering, as Exhibit Four shows. (The bars in the exhibit show the numbers of USCs being awarded science and engineering PhDs, and for the other three degrees the total number of degrees awarded by U.S. institutions. Since few non-USCs receive U.S. MDs, LLBs, or MBAs, and data on them are uneven, we show the total number of degrees granted, with no breakout by civil status.)

The actual number of science and engineering degrees awarded to USCs was down a little in 1990 (10,412) from the 1970 level (11,018) but up a bit from 1980 (8,942). Against this relatively flat profile, we see that the number of medical degrees had increased by one half in twenty years, law degrees had jumped by a factor of two-and-a-half, while business master's degrees had increased three-fold.

In general terms, then, it appears that U.S. college graduates are making economically-rational decisions (for themselves); if they are interested in graduate school of any kind, they tend to move toward the graduate degree programs that lead to the highest incomes.

People with U.S. bachelor's degrees in engineering, however, have more options. They can go directly into well-paid professional work *without* going to graduate school, or they can aim for higher pay (later) by seeking a PhD. The 1992 salary survey of the American Institute of Chemical Engineers casts some light on this decision.[25] What happens when a potential graduate student in chemical engineering applies some simple economics to the salary data?

If one pays attention only to the financial reward for graduate study (a major, but certainly not the only, consideration), at first blush the decision to seek the PhD makes sense. The median salary for all those with doctorates responding to the survey was $67,200 against $54,000 for those with bachelor's only. Further, if we look at the experience of the BS class of 1986, we see that those with

---

[25] The author is grateful to Dr. Bruce C. Robertson for working with him on the passage that follows and on Exhibit Five; Dr. Robertson holds a PhD in Chemical Engineering from the University of Delaware, and is Manager, Catalysis Research, W. R. Grace & Co.

Exhibit Four

Trends in U.S. Graduate Degrees
(1970-1990)

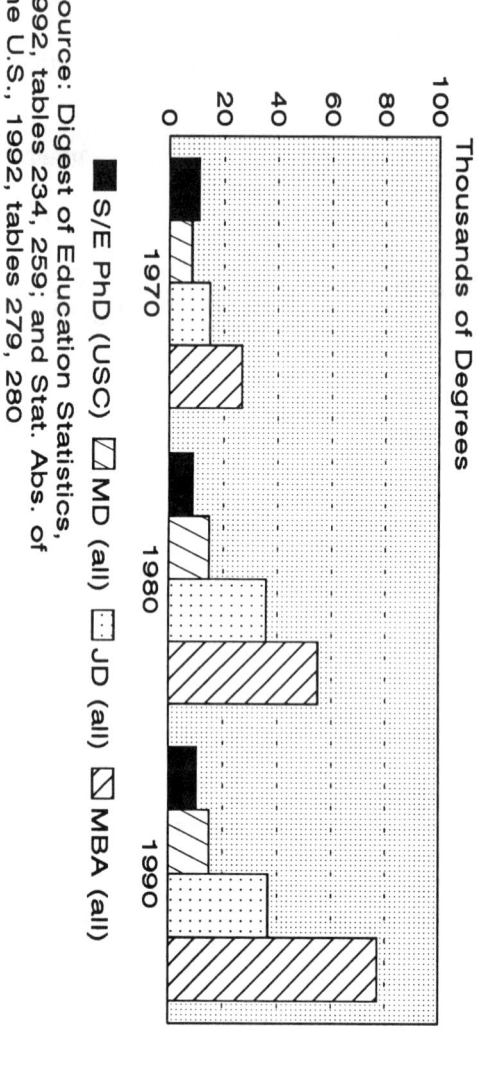

Source: Digest of Education Statistics, 1992, tables 234, 259; and Stat. Abs. of the U.S., 1992, tables 279, 280

bachelor's degrees only received $45,500 against $55,250 for their peers with doctorates.[26]

A closer examination reveals, however, that the rewards for spending those long years in graduate school are less attractive financially. In Exhibit Five we assumed that the current salary status of chemical engineers, those with bachelor's only and those with doctorates, is an appropriate predictor of future salary differentials.

What we find is that the lost earnings during graduate school are substantial enough so that it takes 23 years before the cumulative earnings of the PhD exceed those of the BS chemical engineer. If these data and assumptions are correct (a point to which we will return), the PhD chemical engineer begins to be financially rewarded for the graduate school decision only after reaching the age of 45!

We made a number of assumptions in these calculations. First, we assumed that, on average, chemical engineers work (or study and work) for 42.5 years beyond the BS. With a new BS graduate averaging an assumed 22 years of age, this suggests retirement at an average age of 64.5. (If we are setting retirement too late, we are overstating the advantage of securing the PhD, which provides most of its benefits in the last 17 years of a 42.5 year career.)

We also assumed that grad students paid no tuition and that they secured $14,500 a year in stipends over a 4.5-year period in graduate study. (To the extent that we may have overstated the graduate students' income during these years, we would also have overstated the mild financial advantage to those with doctorates.)

Social scientists call these studies cross-sectional analyses, and sometimes worry about the methodology's utility as a predictor. The alternative methodology would be to conduct a cohort survey, of those who secured their BS (ChE) in, say 1949, following their salary patterns over their lifetimes. This approach also has problems, such as reflecting PhD/BS earnings patterns of the past, rather than current ones; to our knowledge no one has conducted such a survey.

That said, what do the numbers show? Lifetime earnings of those with doctorates exceed those with bachelor's by only $154,050 in pre-tax dollars. That is a margin of less than 6% of total earnings. Further, if income-based taxes (federal, state, and social security) were factored in at, say, a lifetime marginal rate of 40%, the lifetime

---

[26] *Chemical Engineering Progress*, July, 1992, pp 53-62.

Exhibit Five: Differential Earnings Over Time of Bachelor's Only and PhD Chemical Engineers-- Or Why Many Avoid Grad School

Passage of time since Bachelor's	Median Earnings, Bachelor's only	Median Earnings, PhDs	Advantage to those with Bachelor's only
0 - 4.5 years	$180,350	*$65,250	+$115,100
4.5-10 years	$259,000	$300,075	
cum. 10 years	$439,350	$365,325	+$74,025
11-20 years	$607,000	$627,000	
cum. 20 years	$1,046,350	$992,325	+$54,025
21-30 years	$739,550	$807,350	
cum. 30 years	$1,785,900	$1,799,625	-$13,775
31-40 years	$761,250	$856,875	
cum. 40 years	$2,547,150	$2,656,550	-$109,400
41-42.5 years	$183,700	$228,350	
career total	$2,730,850	$2,884,900	-$154,050

* Assumes graduate student receives full tuition scholarship plus $14,500 a year in stipends and assumes 4.5 year period in graduate school, hence $65,250 in income as a grad student. The table also assumes chemical engineers will work for 42.5 years after receiving their bachelor's degree, until about 64.5 years of age. For other assumptions see text.

Source: calculated by the author and Dr. Bruce C. Robertson from AIChE's 1992 salary survey published in *Chemical Engineering Progress*, May 1992, p. 59.

---

difference in dollars would be less than $93,000.

Another significant point is the concept of current value of money, that is, $10,000 received in 1992 is much more valuable than $10,000 to be earned twenty years later. For example, let's look at the $115,100 higher earnings of the BS-only graduates during the first 4.5 years after college (far right column of Exhibit Five). If this money had been invested at a compound interest rate of a modest 5% it would grow, after 40 years, to a thunderous $810,304. And one can, even in times of low interest rates, generally secure tax-free bonds paying more than 5%.

A more realistic way to look at the early lost earnings of PhD chemical engineers would be to apply the current value of money notion to a much smaller sum. Let's say that new ChEs with the BS could save about 10% of their salaries, or about $4,000 a year during the 4.5 years when their peers are in graduate school. That would produce $18,000 and, 40 years later, that would be worth $126,720, close to the $154,050 (pre-tax) difference between the lifetime earnings of the doctorates over the bachelor's. This analysis of the salary data, not unlike others done on this point, suggests that the salaries paid PhD chemical engineers are not large enough to encourage attendance at graduate school.

What holds for chemical engineers, however, is not equally true for chemists. While chemical engineers can become full-fledged professionals without a PhD, this is not always the case in chemistry, where a bachelor's degree often leads to a technician's job. Given this situation, there is a larger gap between the average salaries of chemists with bachelor's degrees and those with doctorates.

For example, here are the median annual earnings of chemists and chemical engineers as recorded in two 1992 salary surveys:

Degree	Chemists	Chemical Engineers[27]
BS	$42,000	$54,000
MS	$50,000	$62,000
PhD	$60,000	$67,200
BS/PhD Difference	+43%	+24%

Why are chemists with the same degree paid less than chemical engineers? Speculation we heard from several sources was that engineers are closer to management than chemists; that engineers

---

[27] The surveys were reported in *Chemical Engineering Progress*, May, 1992, p. 59, and *Chemical and Engineering News*, July 13, 1992, p. 40.

bring the product to market, an activity closer to the corporate bottom line than the more theoretical work of the chemists; and that for these reasons engineers are better paid.

Although PhD chemists earn less than PhD chemical engineers the native born are somewhat more likely to pursue a doctorate in chemistry than in chemical engineering. Of those receiving these two degrees in 1992, 67% of the chemists but only 55% of the chemical engineers were U.S. citizens or resident aliens. [28] The higher doctorate-bachelor's salary differentials in chemistry may encourage more Americans to pursue a PhD in that field.

Graduate School Choices of Foreign Students

Large and growing numbers of overseas college graduates come to the U.S. to pursue science or engineering graduate degrees. Why do they seek these degrees, but not others in the States?

There appear to be four basic reasons:

1. U.S. science and engineering degrees (and the graduate education that they represent) are highly regarded world-wide;

2. Americans are not as interested in these graduate programs, leaving vacancies for well-qualified foreign-born applicants;

3. As we show later, funding, often full-funding, is available to foreign-born students in science and engineering graduate programs, but not in law, medicine, and business; and

4. Many foreign-born applicants regard securing an American PhD as their only way to reach another goal, an immigrant visa to the U.S.[29]

---

[28] Calculated from *Summary Report* 1992, p. 44.

[29] The author is grateful to George Borjas for pointing out the interest in a science or engineering graduate degree as a lever toward a green card. There is yet another motivation for a subpopulation on whom we have less data, the foreign born securing bachelor's degrees in engineering in the U.S. A Sloan Foundation-supported study of the graduate school attitudes of engineering seniors indicated that only 10% of the USC seniors reported a lack of attractive full-time job offers, but that this was

Further, both the science and engineering labor market and the U.S. immigration system make it possible for many of the foreign born with advanced degrees to have their choice — they can stay and work in the United States or they can go home with a valuable vocational credential, the American advanced degree.

Finally, the Indian or Chinese college graduate making this decision is probably in touch with part of the international network of people in his or her field — often through E-mail — and hence is well aware of the availability of graduate school slots, graduate school funding, and U.S. employment possibilities. Although most chain migration is driven through family ties, this network is linked through academic and professional ties.[30]

If able foreign students are entering U.S. science and engineering grad schools in such large numbers, why are they not as visible in law, medical, and business schools?[31] There seem to be two answers:

---

the case with 33% of their foreign-born peers — hence grad school looked more attractive to the foreign born among these seniors. See Elinor G. Barber, Robert P. Morgan and William P. Darby, *Choosing Futures: U.S. and Foreign Student Views of Graduate Engineering Education*, Institute for International Education, New York, 1990, p. 15.

[30] An early NSF study noted that a 1973 survey of S/Es immigrating to the United States in the mid- and late 1960s revealed different motivational patterns by nations of origin. By margins of 60% or more (presumably in response to a question in which multiple answers were permitted), people from Switzerland and Poland said that one of their major reasons for migrating was because they were curious about the U.S.; 98% or so of the Cubans said that they disliked the political environment in their homeland; while the lack of opportunities to do research was a major factor for those from India and China. A variable cited by people from five other countries was a desire for a higher standard of living. National Science Foundation, *Immigrant Scientists and Engineers in the United States: A Study of Characteristics and Attitudes*, (NSF 73-302) USGPO, Washington, D.C., 1973, p. 1.

[31] Given the apparently modest numbers of non-USCs in these institutions, relatively few data are collected on them. The American Association of Medical Colleges asks its new MDs annually if they are USCs. Of the 11,962 responding to the question in 1992, 2.8% answered in the negative (telephone conversation with AAMC). Similarly *Foreign Participation* showed that 1.3% of those securing bachelor's degrees in health fields in 1990 were non-citizens as were 5.4% of those securing master's degrees; see table A-15. Another table in the same report (A-12) indicates that 12.6% of the graduate students in health fields in 1990 were non-USCs. Neither the American Association of Law Schools nor the American Bar Association

first, law, business, and, to a lesser extent, medicine, are country-specific disciplines; and second, it costs considerably less to secure professional degrees in these fields in the home country than it does in the U.S.

While we know the value of a U.S. PhD in science or engineering, we do not have a comparable sense of the value overseas of an American law degree, for example. It probably would not lead easily to a license to practice law, as none of the training would involve the legal principals of the nation in which the new lawyer would practice. What non-USCs are likely to seek, in U.S. law schools, is a different degree, Master of Law or Master of International Law. Usually the degree candidate already has a law degree from a homeland institution, and the U.S. LLM provides a useful credential for a *homeland-based* practice of international commercial law.

Similarly, as U.S. physicians quickly learn when seeking to treat people from a distant culture, there is a lot of cultural specificity in the provision of and acceptance of medical services. It is probably best for physicians to be trained in the milieu where they will work. The terms for symptoms and often the names of medications are language-specific. Further, an in-country medical degree probably leads more quickly to licensure than a foreign degree — it certainly works that way in the U.S.

There are substantial numbers of foreign-born medical doctors in the U.S. but most of them are foreign-trained as well. It is the locus of the training, and some questions about its quality, *not* the place of birth, that has created the interest in the FMGs (Foreign Medical Graduates).[32] In what little we have read about this situation, we have gathered that there is much more concern about the larger population of *foreign-trained* USC physicians than there is about the smaller number of US-trained *foreign-born* physicians.

Similarly, apparently the powerful lure of the MBA has not

---

collects data on noncitizen law graduates.

[32] See, for example, Division of Manpower Intelligence, U.S. Public Health Service *Foreign Medical Graduates and Physician Manpower in the United States*, U.S. Department of HEW, Washington, (DHEW Publication No. (HRA) 74-30), 1974, pp. 48-49.

reached much beyond our shores. The bulk of what is taught in these schools relates directly to the American business culture.

Then there is the matter of money. While an overseas college graduate with good grades can usually secure a free graduate

---

Individual Decisions

While fundamental considerations of long-term career opportunities drive most graduate school decisions, there is always the individual angle. The following three stories were elicited at a convention of the American Chemical Society.

Love: A European, he fell in love with a woman from Milwaukee, and they decided it would be easier for them to do their studying in the United States than in Europe. She could work part time in the U.S. while attending college, but probably could not do so in Europe. (He was fully-funded in either place.)

The Children: "You can raise your children on a U.S. stipend but you cannot on a British one, so I came here. "He added that the U.K. will support a PhD student only for three years, while the period of support is longer here.

Field of Study: A Frenchman, he wanted a PhD in chemistry, but with a master's in geology he could be admitted to a PhD program only in that field in France. He came to the U.S., where he could study what he wanted to study.

---

education in law, medicine, business, science, or engineering in the homeland, the only strong chance for a free graduate education in the U.S. is in science or engineering. Regardless of citizenship or place of birth, it is difficult for anyone to secure full funding for law, medical, or business school in the States; family money or extensive loans are usually needed.

These then are some of the reasons why foreign-born talent flocks to U.S. science and engineering graduate programs, but not to law, business, or medical schools.

The Different Levels of Interest in the Life Sciences

Foreign-born students are not spread evenly through science and engineering graduate programs. A look at some of the areas of concentration shows what percentage of total PhDs were granted to nonimmigrants in 1991:

Discipline	Per Cent Nonimmigrants
Civil Engineering	63.5%
Mechanical Engineering	56.5%
ENGINEERING	51.2%
MATHEMATICS	49.3%
COMPUTER SCIENCE	43.3%
Biometrics and Biostatistics	40.3%
PHYSICS	40.3%
Chemical Engineering	39.8%
CHEMISTRY	31.7%
Biochemistry	26.0%
BIOLOGICAL SCIENCE	22.9%
Geology	17.6%
Zoology	8.2%

Major fields of study are shown above in all capital letters, and subfields in capital and lower case. Note the minimal participation of non-USC graduate students in the descriptive studies (zoology and geology) and the much higher non-USC participation in the one biological subfield (biometrics and biostatistics) stressing statistics.[33]

As to participation in life sciences compared to physical sciences, by various American populations, these data on the 1991 cohort of science and engineering PhDs are instructive:

Population	% of Physical Science Doctorates	% of Life Science Doctorates[34]
USC	57.1%	69.6%
Women	18.4%	38.6%
Blacks	0.8%	1.7%
Hispanics	1.6%	1.8%

One of the principal theories about why Americans generally, as well as women, Blacks, and (perhaps) Hispanics, favor the life

---

[33] Calculated among those with known citizenship from *Summary Report 1991*, pp. 70-71.

[34] For USCs: calculated among those with known citizenship from *Summary Report 1991*, table 5; for others, from table A-2.

sciences as opposed to physical sciences and engineering is the lower emphasis on mathematics in the former. It is often said that people who want a graduate degree in science and engineering but who are not confident of their mathematical abilities will choose a branch of biology. The fear of math is said to be stronger among Americans than among foreign-born PhD candidates. Further, American society has traditionally not encouraged women, or Blacks and Hispanics, to excel at mathematics.

Discussions of this subject, as the previous paragraph illustrates, quickly becomes a matter of generalizations and anecdotes; we encountered nothing more tangible, not even the results of any surveys on the subject. This is unfortunate because there may be something in how the U.S. approaches the life sciences that might suggest ways to approach other sciences, if we want more Americans to study them at advanced levels.

We ran into four other theories about the differential interest in life and physical sciences among the foreign born and the native born:

1. There is a stronger usage of English in life sciences than in other science and engineering fields; this discourages many of the foreign born, but not Americans.

2. The undergraduate training prowess of Asian nations is stronger in the other science and engineering fields than it is in life sciences.

3. While chemistry and physics are rarely encountered in U.S. school systems until high school, U.S. children from the early grades on are exposed to biology; hence, the argument goes, they are more receptive to studying it at the PhD level.

4. Biology was a socially-acceptable study for women even in the 19th Century, on the grounds that women's domain was the home and the garden, so there is a much older tradition of women

working in this field.[35]

With a sense of the characteristics of the foreign-born S/Es, and their motivations, we next turn to the gatekeepers, the organizations that decide whether or not the foreign born can come to, or stay in, the U.S.

---

[35] There is not only a higher percentage of women in biology than in the other hard sciences, but there are differential — and apparently traditional — distributions of women *within* biology as well, according to J.H.M. Willison, Chairman of the Biology Department at Dalhousie University (Halifax, Nova Scotia). He explained that there have been outstanding women biologists for generations and that their current distribution reflects, to a degree, the subspecialty distribution of prior generations of women in the field. In Victorian days, women interested in science were steered away from the math-driven sciences toward biology, and within biology away from subjects dealing with sex, hence towards botany and away from zoology. Within botany there is and has been a strong pattern of women concentrating on algae; why algae? Because it was socially acceptable in days gone by for women to study cryptogams (life forms whose reproductive system are hidden or asexual, such as algae).

Chapter Three

The Gatekeepers

A foreign-born scientist or engineer, wanting to come to the U.S. to secure graduate training and a job along what we call the immigrant path, routinely faces four sets of gatekeepers:

- The Educational Testing Service (ETS) which administers several examinations;

- Graduate school admissions officials;

- Employer decision-makers, and

- The U.S. immigration system.

Of these, ETS is the most nuanced, but rarely makes a yes/no decision about an applicant; its role is to help colleges and universities make decisions about individual applicants for admission.

Universities and corporations *do* make yes/no decisions about individual foreign-born S/Es, as they seek graduate school admission or professional employment.

The immigration system, with rare exceptions, plays a passive role, routinely ratifying decisions made by the universities and corporations.

The Educational Testing Service

ETS is a large nonprofit organization located near Princeton,

New Jersey; it was established by America's universities to create and administer standardized tests for the admissions process. At first there were just the College Boards, used to decide admissions at the undergraduate level; now there is a wide array of tests, used by many different U.S. and Canadian educational institutions.[36] Of particular interest for the purposes of this study are the Graduate Record Examination (GRE) and the Test of English as a Foreign Language (TOEFL).

The GRE is routinely used by admissions officials at U.S. graduate schools (other than those in law and medicine) to supplement other information they secure, such as undergraduate grade-point averages and recommendations. The test is given several times a year at locations all over the globe. The applicants, under the watchful eye of a proctor, complete a multiple-choice examination, which takes about five hours.

The GRE is divided into three sections, testing different kinds of abilities: verbal, quantitative, and analytical. The first tests how well one reads and understands English and the size of one's vocabulary; the next deals with mathematical skills, but does not assume detailed knowledge of college-level math; and the third tests analytical reasoning (through questions asked in English). As a participant-observer, the author took the GRE two years ago, and came away quite impressed with the high scores in the second and third parts of the test that many applicants routinely achieve; the author's results were not so impressive. (It *had* been more than 40 years since the author's last math course.) The scoring range on each of the three parts of the test is from 200 (which you get for merely signing your name) to 800 for a perfect paper.

Graduate school applicants from non-English speaking countries take the TOEFL as well as the GRE. TOEFL, also given around the globe, is a multiple-choice test designed to gauge one's ability to understand English as spoken in North America and to read it. It tests neither the ability to *write* English nor to *speak* it; there are ETS tests for both those skills but, given their expensive nature (resulting from the labor-intensive scoring involved) neither is currently used in

---

[36] Although ETS conducts its examinations worldwide, the use of the scores is largely confined to North America; educational institutions in other nations use other systems for making admissions decisions.

the graduate school admissions process. TOEFL scores range from 200 to 677.

Some controversy surrounds both the GRE and the TOEFL. The GRE, along with other ETS systems, has been attacked largely from the left as being culturally-biased (tilted towards middle class white Americans),[37] but foreign-born S/Es do not seem to be making this argument, perhaps because the ones who do not do well in the GRE are not in the U.S. to protest. We conclude, however, that (1) there is bound to be some cultural bias in any standardized test despite the best efforts of the writers[38]; and (2) if there is a significant cultural bias it would tend to be less fair to non-U.S. participants than it would be to U.S. minorities.

Another criticism of the GRE, a centrist one, heard from some U.S. research universities, is that it is not nuanced enough to sort through the most able of the applicants effectively. "What good is the GRE to us if virtually all of the candidates score 800 on their math test?" was a question raised at one of these institutions. (News reports indicate that ETS has developed a new version of the GRE, to be taken on a personal computer; the new version is designed to take care of this problem.)

Meanwhile, a third criticism, this one of the TOEFL, from what could be regarded as the right, asserts that the test is too easy, and that it allows the admission of graduate students with limited English ability who subsequently serve as teachers' assistants (TAs). The argument is that the poor English of some science and engineering

---

[37] For example, Frank Morris, Dean of the Graduate School of Morgan State University, who is generally critical of the way American Black males are treated vis-a-vis graduate training in science and engineering, objects to what he regards as the over-use of standardized tests in the graduate school admissions process; interview with Dr. Morris, September 24, 1992.

[38] The author has no first-hand contact with the development of the GRE but did experience the remarkably painstaking efforts of his wife and her colleagues in the construction of questions for the Law School Aptitude Test (LSAT). Each question went through multiple levels of tough internal review before it became a tentative part of LSAT. The new question was then tried with a round of LSAT test-takers; results were not included in applicants' test scores but were tested against the balance of the examination. If the trial questions turned out to produce comparable scores to the rest of the examination, they could be used as functional parts of the LSAT in subsequent iterations.

TAs discourages Americans from continuing to study these subjects.[39]

> **The Senator's Daughter**
>
> She told her father that the reason she failed the test was because she could not understand the TA's English. Daddy was a member of the Texas State Senate. Soon the appropriations bill for the University of Texas mandated that TAs be tested for verbal skills in English and be tutored if they lacked them.

Graduate Record Examination

While ETS wrestles with these criticisms, the tests go on, in large and increasing numbers. In 1991, for example, 450,000 GREs were conducted, including about 100,000 overseas. Of the total, about 150,000 were candidates for PhDs;[40] since only about 37,000 PhDs are awarded annually in the U.S. this would suggest a four-to-one ratio between would-be and actual doctorate holders.

How well do American and overseas applicants do on the GRE? Although it is American-designed, Americans do not sweep the field. As Exhibit Six shows, the six nations listed in Asia (as well as Canada and England) have higher scores than the U.S. on the quantitative

---

[39] Two Sloan-funded studies have touched upon this subject. One of these studies is Nancy M. Hewitt and Elaine Seymour "Factors Contributing to High Attrition Rates Among Science, Mathematics and Engineering Undergraduate Majors," Bureau of Sociological Research, University of Colorado, Boulder, 1991. The foreign-born TAs are not mentioned in the report, but Dr. Seymour told me that while some of the students spoke of this factor, none said that it caused them to leave the studied majors. George Borjas, at UC/San Diego, however, found "a strong negative impact of foreign-born teaching assistants (who are used commonly not only in economics but in other fields as well) on the class performance of American undergraduates. On average, undergraduates enrolled in sections taught by a foreign-born TA can expect to receive a grade that is about .2 grade points lower (or about the difference between a B+ and a B) than undergraduates taught by native-born TAs, even after controlling for a student's innate ability." See George J. Borjas "Do Foreign-Born Graduate Students Affect the Scholastic Achievement of Undergraduates?" mimeo, Sloan Foundation, November 1994.

[40] Interview with Charlotte Kuh, a senior ETS staff member, June 5, 1992.

Exhibit Six: GRE General Test Scores for Selected Populations: 1987-1988
(With Rankings of Four Top Groups in Each Category)

Population	Number	Quantitative	Analytical	Verbal
U.S. Applicants	203,084	537	542 (3)	508 (3)
Overseas Applicants	72,806	617	480	401
Taiwan	4,534	727 (1)	525	341
Hong Kong	1,099	698 (2)	537 (4)	376
China (PRC)	5,544	686 (3)	467	347
Korea	2,932	683 (4)	431	375
Japan	1,783	662	470	331
India	9,383	652	519	468 (4)
England	871	600	583 (1)	573 (1)
Canada	2,876	580	562 (2)	537 (2)
Mexico	757	521	448	399
Mexican-Americans, US	2,993	461	461	444
Blacks, US	12,592	394	406	391

Note: These are the mean scores received by these populations of would-be graduate students after taking the paper-and-pencil GRE test; the scoring range is from 200 (for answering no questions) to 800 (for a perfect paper). The populations shown are three U.S. groupings, the seven overseas nations with the largest number of applicants (in descending order: India, PRC, Taiwan, Korea, Canada, Japan, and Hong Kong) and Mexico and England. Populations are arrayed according to their quantitative test scores.

Source: Diane M. Wah and Dawn S. Robinson, *Examinee and Score Trends for the GRE General Test: 1977-78, 1982-83, 1986-87 and 1987-88*, Educational Testing Service, Princeton, N.J., 1990, Tables 2.2 and 2.3.

examination. Taiwanese students average almost 200 points higher than U.S. students. And Brits and Canadians outscore Americans on the verbal and analytical segments. U.S. Blacks and Mexican-Americans have low scores. This exhibit covers would-be graduate students in the arts, sciences, engineering, and humanities.

When we narrow the focus and look only at GRE examinees seeking to study the physical sciences (including mathematics and computer science), engineering, or the life sciences, we find that the foreign-born test takers do better on the quantitative examination than Americans, but not quite so dramatically (see Exhibit Seven).

In each of the three categories, most notably in life sciences, the overseas examinees did better than U.S. citizens in the quantitative test but less well in the other two tests. And in every discipline, and in every test, coming in last were the immigrants to the U.S. (noted as PRA in the exhibit).

Not shown in either exhibit is the clustering of the overseas GRE candidates at the very top category of the quantitative test, those who solved every problem correctly and thus secured a score of 800. More than 12% of the nonimmigrant test-takers seeking admission to graduate school in the physical sciences scored 800, compared to 5% of the U.S. citizens and 4% of the immigrants. In engineering more than 9% of the nonimmigrants scored 800, compared to a little less than 5% of the U.S. citizens. The percentages were lower, but the relations were similar for those seeking admission to graduate school in life sciences. As Charlotte Kuh, a senior ETS official who works with the GRE, wrote to the author:

> I want to add a quick note about drawing inferences from these score distributions. Although people in the sciences, especially those who are not U.S. citizens, do exceedingly well on the GRE Quantitative Reasoning measure, you should note that a very high percentage of these people score above 700. Rather than indicating that these people are wonderfully gifted in math, this distribution of scores indicates that the GRE Quantitative measure is not particularly appropriate to this population. The context in which it tests math reasoning does not extend beyond high school algebra and geometry. We are currently developing a measure that will be more suitable for this group, but we don't have one right now.[41]

---

[41] Letter to author, April 18, 1994.

Exhibit Seven: General GRE Mean Test Scores for Would-be Graduate Students in Science and Engineering, 1993

Discipline	GRE Test	USC	PRA	Nonimmi-grants	USC/ Nonim. diff.
Physical Sc., Math & Comput.	Quantitative	656	643	695	-39
"	Analytical	614	495	552	+62
"	Verbal	529	398	431	+98
Test takers		12,776	1,044	8,261	
Engineering	Quantitative	679	651	699	-20
"	Analytical	608	488	536	+72
"	Verbal	513	392	430	+83
Test Takers		15,570	1,553	14,045	
Life Sciences	Quantitative	574	554	622	-48
"	Analytical	577	475	503	+74
"	Verbal	508	421	424	+84
Test Takers		16,458	735	4,894	

Key: USC = United States Citizen; PRA = permanent resident alien (i.e. an alien in immigrant status); Nonimmigrant = an alien who is neither a USC nor a PRA, and, if admitted to the U.S., would probably have a student or exchange visa.

Source: Unpublished data from the Educational Testing Service, Princeton, N.J.

Despite these cautionary words and despite the knowledge that many other factors are considered in addition to the GRE Quantitative examination, we conducted a small experiment based on the 1993 results. What would the graduate school class that year look like, in terms of citizens, immigrants, and nonimmigrants, if admissions were based only on the Quantitative examination?

We made three other assumptions. First, that the PhD programs would accept about as many new graduate students in the fall of 1993 as secured PhDs in the previous year. Second, that all the top-scoring applicants were seeking PhDs rather than master's. Third, that all candidates would accept the offers of admission.

With all this in mind, we were looking for at least 6,498 candidates in the physical sciences, 5,437 in engineering, and 7,108 in the life sciences. We worked with the slightly higher totals shown below, as GRE scores are reported in units of ten. In each instance we set the cutting-edge score at the highest level that would fill the vacancies. To fill physical science slots, using only Quantitative GRE scores, we would have to go only to down to the 750 level. Those accepted would be distributed as follows:

	780-800 Scores	750-770	Total[42]
Citizens	1,658	1,418	3,076
Nonimmigrants	2,083	1,260	3,343
Resident aliens	123	99	222
All admitted	3,864	2,777	6,641

Note that an absolute majority of the slots would go to nonimmigrants and that the percentage of nonimmigrants would be higher among the first group to be selected (780-800) than the second group, those with slightly lower scores.

Going through the same exercise with would-be graduate students in engineering, we would stay above 780 and admit 3,053 nonimmigrants, 156 immigrants, and 2,056 citizens (less than 40% of the cohort). With the life sciences, the cutting edge score would drop to 650, and we would admit 4,624 citizens, 197 immigrants, and 2,301 nonimmigrants. The ratios of nonimmigrants to citizens in these

---

[42] Source of data: unpublished tabulations of 1993 GRE scores from ETS.

three exercises are higher than they are in actual graduate school enrollments, a subject to be covered later.

In addition to their comments on the appropriateness of the examination, ETS officials suggest that there are at least two other factors playing roles in the results just discussed. First, ETS believes that would-be graduate students overseas are much more likely to train for the GRE than Americans are. Second, a much more elite group takes the GRE overseas than in the U.S. About one in every thousand Americans takes the test each year; in 1987 only one resident of PRC in 200,000 took the test.

While ETS and others say that too much emphasis should not be placed on a single set of examination results, it seems that nonimmigrants seeking advanced U.S. degrees in science and engineering are, at the very least, better at taking the Quantitative GRE test than are their American peers.

Test of English as a Foreign Language

ETS also handles more than 550,000 TOEFL examinations each year, though most of these are not for graduate admissions.

Given the intense interest in admission to U.S. grad schools and the small number of slots compared to the large number of applicants, there is always the threat of fraud, both overseas and in the U.S. ETS is aggressive on this point and has never lost a court case for refusing to honor a questionable test score.

The threat scenario of most concern to ETS is the substitution of test-takers in the TOEFL examination.[43] With this in mind, the TOEFL score is conveyed to the North American university on a highly secure piece of paper bearing not only the applicant's name, address, signature, and passport number, but a photograph that was accepted at the test site as matching the face of the applicant *and* the photo in the passport presented at that time. Further, the photograph is computer-printed on the score sheet. When the applicant arrives at the North American university his or her appearance and documentation can both be checked against those

---

[43] GRE scores, university grades, and professors' recommendations should all paint a roughly similar picture of a student's academic prowess; there is usually no independent check — beyond TOEFL — of an applicant's ability with English.

shown in the TOEFL results. Further, if one tries to photocopy a TOEFL document the word "copy" appears all over the reproduction. This system is more sophisticated and more secure than anything the U.S. government does to make sure that an alien arriving in the U.S. is the same person to whom the entry documents had been issued in the first place.

Graduate School Admissions[44]

Graduate school admissions processes are totally different from those used at the undergraduate level. Undergraduate decisions are made *en masse* as the institution in a centralized process seeks to "build a class" to fit the needs of the university. Not only should it consist of able students, but it also should have the appropriate mix of men and women, of majority and minority populations, and of American and foreign students. Further, to some extent adjustments are made for special university interests, such as admissions of athletes and children of alumni. All these factors, as well as Scholastic Aptitude Tests (SATs), grade point averages, teachers' recommendations, and, in many cases, interviews, are used as the university decides whom to admit at the undergraduate level.

In science and engineering grad schools it is different. The criteria are almost exclusively those of a meritocracy, and the decision-making is fragmented among the departments. Further, the graduate school admissions process deals almost exclusively with paper. Rarely are applicants interviewed. Some reliance is placed on grade-point average and on GRE scores, but more is placed on the graduating rank of applicants from institutions known to the U.S. graduate department. If the Harvard Department of Chemistry

---

[44] This section is based in part on conversations with numerous university officials and faculty members, including: Assistant Deans Marianne Waterbury and Joy Montero of Princeton University, Ms. Geri Rypema of George Washington University, Professor Andrew Zydney (Chemical Engineering) of the University of Delaware, Professor Eugene Skolnicoff (Political Science) and Associate Dean J. Jackson of MIT, Professor Peter Chen (Chemistry) and Ms. Sharon Ladd of Harvard University, Professor J.H.M. Willison (Biology), Ms. Barbara Maynard, and Ms. Ann Thomson of Dalhousie University, Dr. Jules B. LaPidus, President of the Council of Graduate Schools, and many others. There also were extensive conversations with science and engineering graduate students and recent foreign-born and native-born recipients of science and engineering doctorates.

knows the Chemistry Department at a particular university in India, it is much more likely to accept a candidate who did well at that institution than one who did equally well at an unknown overseas university. The old-boy network plays a role as well; a strong recommendation from Professor Y at University X to a professor he knows at an American university can be crucial.

In some of the larger departments, such as electrical engineering at MIT, these overseas contacts are handled on a highly organized basis. Several senior members of the faculty have each been assigned a different part of the world and have been encouraged to become familiar with the institutions there that produce bachelor's of electrical engineering. These professors are then given the task of initially reviewing the applications coming from those nations.[45]

The graduate admissions officer for the university usually plays only a minor role in this decision-making process, for example, making sure that the departments have followed university regulations and filled out the forms correctly. Further, in more affluent institutions where all graduate students are routinely fully funded, the admissions officer makes sure that the departments have the funds needed to cover the number of people admitted.

Sometimes the decision-making process is decentralized below the departmental level; we heard of graduate students seeking admission so that they could work under a specific professor. That professor then decides if he or she wants the student, and if the research budget can support the admission; if both answers are "yes" the student is likely to be admitted. Foreign-born graduate students, however, are less likely to follow this route than American ones, usually because the foreign born do not have that much access to academic networks prior to their arrival. Once the department has made its decision, the university nearly always supports the department, and the INS nearly always supports the university.

Within the process described above, and despite the growing numbers and percentage of foreign-born science and engineering graduate students, there is a certain preference shown for citizen students, by both U.S. and Canadian institutions. This was stated clearly and openly in many of my conversations with academics and university officials. In some cases it is a mild, general preference; in

---

[45] Interview with Professor Emeritus Louis Smullin, MIT, May 23, 1993.

others there is a well-established ceiling of, say, 30% to 50% for the number of foreign applicants accepted in a given class, though we never saw anything in writing on this point.

It is the stronger institutions, those attracting relatively large numbers of qualified applicants, that can set and enforce these ceilings. Less highly regarded institutions tend to take on larger percentages of foreign-born graduate students. For example, in Goodwin and Nacht's perceptive report, this example is cited:

> Officials of one institution of relatively low national prestige admitted that they had used foreign students as "filler" in the growth process of the institution. In the early days of certain graduate and professional programs, they were simply unable to attract the required number of U.S. students essential for faculty growth — and so they turned overseas. When the programs achieved a reputation, they were able to replace the foreigners with U.S. students.[46]

Some quantitative data on this highly qualitative matter appeared in one of the NSF publications. Using a standard measure of departmental quality, the reputational survey of the Conference Boards of Associated Research Councils, the NSF researchers cross-tabulated the percentage of foreign-born graduate students in 1983 with the qualitative rankings, with these results:

Quality of Department	Percentage of Foreign-Born Graduate Students			
	Eng.	Math.	Phys. Sci.	Life[47] Sci.
Highest qtr	37.2	37.1	21.3	11.8
Second	44.5	39.0	23.8	16.8
Third	47.5	36.5	25.5	14.3
Lowest	50.6	38.8	29.2	15.8

---

[46] Crauford D. Goodwin and Michael Nacht, *Absence of Decision: Foreign Students in American Colleges and Universities*, Institute for International Education, New York, 1983, p. 14. The report is subtitled: "A report on policy formation and the lack thereof."

[47] National Science Foundation, *Foreign Citizens in U.S. Science and Engineering: History, Status and Outlook*, (NSF 86-305 Revised), NSF, Washington, D.C., 1987, pp. 19-21 and Table B-20.

Statistical documentation of the widespread preference for citizen PhD candidates can also be found in a study sponsored by the Association of American Universities and the Association of Graduate Schools. The study covered admissions decisions made during the fall of 1990 in 43 graduate institutions providing PhD programs in biochemistry, economics, English, mathematics, and mechanical engineering.[48] These were among the conclusions of that study:

> ...in four of the fields studied (i.e., all except English), non-U.S. citizens constitute a sizeable majority of applicants to doctoral programs....For economics, mathematics, and mechanical engineering, the proportion of enrollees who are U.S. citizens is substantially greater than the proportion of U.S. citizen admittees, which in turn is substantially greater than the proportion of U.S. citizen applicants. For biochemistry, the proportion of U.S. citizen admittees is substantially greater than the proportion of U.S. citizen applicants....Thus, except for English, which has a very large percentage of U.S. citizen applicants, the participating institutions have a clear preference for U.S. citizen applicants. It follows, therefore, that the large proportion of non-U.S. citizen enrollment in these four fields (42 percent overall) is the consequence of the relatively small pool of U.S. citizen applicants, and *not* the result of any bias by doctoral programs for non-U.S. students.

(Emphasis in the original.)

In other words, if you were a noncitizen applicant to graduate school, in these disciplines, at these 43 institutions, you were less likely to be admitted and less likely to be subsequently enrolled than a U.S. citizen applicant. (See Exhibit Eight.)

The study also compared admissions experiences of the large number of foreign applicants and the small numbers of U.S. Black and Hispanic applicants:

> ...both the acceptance and enrollment rates of minority applicants are significantly higher in comparison to those of non-U.S. citizen applicants....[T]his finding does suggest that institutions do show a preference for admitting U.S. minority applicants rather than non-U.S.

---

[48] "Participation in Doctoral Education at Major Research Universities by U.S. Citizens, Women, and Underrepresented Minorities," *Program Profiles*, Vol. 1, No. 1, April 1993, AAU/AGS Project on Doctoral Education, then located at University of Rochester now located at Educational Testing Service in Princeton, N.J.

> citizen applicants. Nevertheless...individuals of minority status are *not* pursuing doctoral education in proportion to their share in the overall population.... [O]f 1,008 U.S. citizen biochemistry applicants, only 18 were Black Americans and only 19 were Hispanic Americans.

(Emphasis in the original.)

In short, if one looks at this situation at one specific but significant point along the spectrum — after people have applied to graduate school — discrimination against Blacks and Hispanics does not occur. The problem is that too few of them have secured the necessary undergraduate education and/or motivation to apply for science or engineering graduate school. The failing then, is in the education system more broadly, one could argue, rather than in the graduate admissions process.

There is another difference between undergraduate and graduate programs at the same set of universities. Often the undergraduate admissions process appears to use affirmative action programs to bring minorities to the campus. Further, there are programs to help those minority students needing help to complete their degrees.

These policies and those helping offices do not seem to exist in graduate schools; despite the admissions preference for citizens, we encountered no special programs in graduate schools designed to help citizens complete their PhDs.

Getting A Job in the U.S.

Once the foreign-born scientist or engineer is near the end of his or her degree program, he or she faces a number of inter-related decisions:

- Do I want to stay in the U.S.?
- If I do, can I get a job in the U.S.?
- Do I want to work in industry or at a university?
- What can I do about my immigration status?

The last question is, of course, of no concern to the individual who has decided to return to the homeland or to work for an

Exhibit Eight: Scatterplot of 142 PhD Programs in Biochemistry, Economics, Mathematics, and Mechanical Engineering at 43 U.S. Institutions According to the Percent of Applicants and of Admitted Applicants who are U.S. Citizens

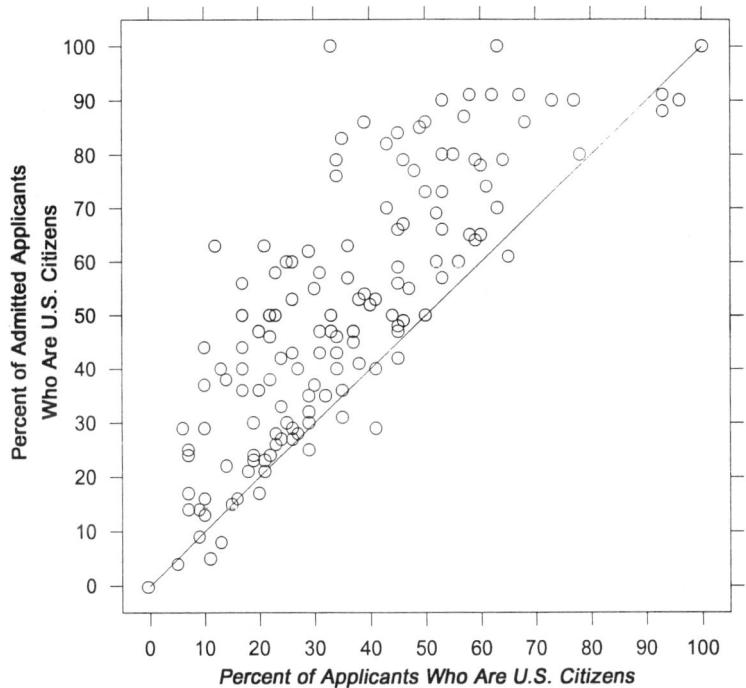

Source: Association of American Universities/Association of Graduate Schools Project for Research on Doctoral Education (1991, September), *Major Research Universities Endeavoring to Increase the Number of U.S. Citizens Enrolled in Ph.D. Programs in Science and Engineering: Efforts hampered by low number of U.S. applicants*, (Research Bulletin 91-01) Rochester, NY: AAU/AGS Project for Research on Doctoral Education, University of Rochester. The study was based on graduate school admissions data obtained from 43 U.S. universities.

international agency, nor to the few who have already secured green card status (usually through marriage), but it is important to most of the new degree holders. And while, as suggested earlier, the immigration system rarely says no to an American institution wanting to educate or hire a foreign-born scientist or engineer, the immigration system plays a complicating role in the job-seeking process.

In very general terms, there are three types of responses (on the immigration variable) from possible employers. Some of the smaller ones are worried about the immigration process and will not consider job candidates who do not already have the green card. Other firms, interested in and possibly already employing foreign-born talent, are willing to hire nonimmigrant S/Es; they will sign any needed papers but will not take care of the process. Their attitude is if there are to be lawyers' fees, let the worker pay them. Finally, there are the larger firms that regard the expense (and nuisance) of handling the worker's immigration papers as a normal cost of doing business. (This is also the usual reaction of universities.)

Corporations typically use attorneys to complete INS petitions; this work is done at universities, at considerably lower cost, by foreign student advisers. We heard from several sources that corporations routinely pay their attorneys $1500-$2000 or so for the nonimmigrant petition and from $5,000 to $15,000 for the more complex (and more enduring) green card process. The latter usually includes filing a labor certification with the U.S. Department of Labor (DOL) and, after this has been approved, a petition with INS. Were the corporations to take their immigration business away from the big law firms that they typically use (the K Street or Wall Street firms) and turn to equally skilled immigration practitioners who work alone, they could cut the green card legal costs substantially.

Some employers in the third category — those paying attorneys' fees — deliberately go through a preliminary stage in which the science or engineering worker spends his or her first six months or year in a nonimmigrant status before the employer takes the next step, providing the worker with a green card. Sometimes the job negotiation includes this variable, with the applicant trying to get a green card included as part of the total employment package.

As a matter of fact, there was a vigorous discussion of this subject at the 1992 meeting of the National Association for Foreign Student Affairs: Association of International Educators (NAFSA), held in

Chicago. An official of one of the major research universities said that *his* institution would not offer a green card to anyone except a member of the faculty on a tenure track; he was unhappy with (apparently lesser) institutions that would offer a green card to untenured faculty and even to postdocs.[49] The most interesting part of these remarks was what was left unsaid — everyone understood that these were decisions to be made by the universities themselves, not by INS.

The way in which corporate employers decide to hire scientists and engineers is quite different from the graduate school admissions process in that there is much less reliance on paper and much more on personal contact and references. Corporations seem more concerned about how the new employee will get along with his or her colleagues than do universities (in the graduate admissions process). Further, the corporations have more money to explore this variable than universities typically have.

Corporate recruiters, often scientists and engineers who work in the mainstream of the company's activities, as well as personnel specialists, visit universities and talk with people about to secure degrees of interest to the corporation. Sometimes the interviewed graduate student or the student's faculty mentor is already known to the corporation as someone doing interesting work in a field related to the firm's activities. If the recruiter is impressed and the corporate staff has an interest in a candidate, an offer is made to bring the candidate to the laboratory or office concerned. Some mutual selling follows, with the graduate student stressing his or her abilities and the corporate staff emphasizing the virtues of the corporation and its location. The latter can be important; a big-city person may not want to move to a largely rural area, or someone with small children may not want them exposed to adverse environmental conditions, such as those in Nitro, West Virginia, or Baton Rouge, Louisiana.

The presence or absence of other people from the homeland in the employer's location may also play a role in the decision by the

---

[49] The author was one of thousands of people attending this conference held over the Memorial Day weekend in a cluster of Chicago hotels. It speaks well of the membership that they would give up a holiday weekend for such a meeting; the timing, of course, reduced the cost of the conferences, on two counts: big city hotels, usually deserted on such weekends, lower their rates, and all air travelers were able to use the over-Saturday-night discounts.

foreign-born scientist or engineer, but we have no data on that point. It is well known as applied to immigrants generally that networks of friends from hometowns (or more likely, in the case of S/Es, universities) help steer people into jobs, and these networks must be active among foreign-born PhDs as well.

Are foreign-born scientists and engineers discriminated against in the corporate hiring process? Nearly twenty years ago, a former corporate recruiter (and PhD candidate himself) wrote his dissertation on this subject and found extensive, if discreet, discrimination against foreign-born S/Es not only within his corporation but within others as well.[50] But that was at a time when perhaps nativism was stronger, and certainly was at a time when foreign-born S/Es constituted a much smaller portion of the work force. Despite extensive probing we found little evidence or even anecdotes on this point. As to the possibility of *post-hiring* discrimination within the corporate world, such as lower salaries and the existence of a glass ceiling, see Chapter 4.

A substantial portion of new foreign-born PhDs in science and engineering never leave academe. They move into positions as postdocs and later as nonfaculty researchers, or into faculty positions. As they move in this direction, they encounter another academic decision-making process, more like corporate hiring than the graduate admissions process, in that interviews and personal contacts play a major role.

Faculty hiring, like the graduate school admissions process, is also highly decentralized. Decisions on hiring post-docs, junior faculty, and researchers are made at the departmental level, and unless something is remarkably amiss, ratified at the university level. Bringing in a new chairman of a department from outside the university, however, involves the university leadership.

Many universities, in effect, mandate internal migration within the U.S. as part of their professional personnel policies. It is a good idea, they say, for one to secure a doctorate from a different institution than where one secured a bachelor's; further, it is a good idea to do one's postdoc work at a different place than the one

---

[50] Bradley W. Parlin, *Immigrant Professionals in the United States: Discrimination in the Scientific Labor Market*, Praeger, New York, 1976. We sought to discuss with Professor Parlin his current reaction to this matter, but were unable to reach him.

awarding the doctorate, and to move on to the faculty at yet another institution. Some institutions, such as Harvard's Chemistry Department, tend to carry this near-mandatory migration one step further — one is more likely to make it to the top of the heap, a tenured full-professorship, by doing well at another institution and then by being invited to a tenured position at Harvard.

Thus foreign-born S/Es who thought they made their migration decisions in their early twenties (when applying to a U.S. graduate school) often find that this was just the first of many such decisions.

The Immigration Process

Three federal departments are involved in the U.S. immigration process. The lead agency is INS, a part of the U.S. Department of Justice. Also playing roles are the Employment and Training Administration of the U.S. Department of Labor (DOL) and the consular segment of the U.S. Department of State.

A typical foreign-born scientist or engineer would have the least contact with the State Department. Once accepted by a U.S. university, the student takes his or her acceptance papers to a U.S. consular office overseas and secures a nonimmigrant visa to come to the U.S. Many years later, probably ten or so, the foreign-born scientist or engineer who had successfully secured a graduate degree, a job, and his or her naturalization papers would come back to the State Department again to get a U.S. passport. There probably would be no contact in the intervening period.

The role of the U.S. Department of Labor relates to wages and working conditions; employers of certain classes of nonimmigrant and immigrant workers are supposed to show DOL that their proposed employment will not adversely affect wages and working conditions of similarly employed U.S. workers. In recent years this has been an area of noncontroversy between DOL and the employers of foreign-born S/Es. This process puts a typical foreign-born S/E, or more likely his or her employer, in contact with DOL once or sometimes twice in the individual's career.[51]

---

[51] The author served as Assistant to the Secretary of Labor, specializing in foreign worker matters, between 1965 and 1967. This was the period immediately following the repeal of the bracero program, which had involved the annual migration

Although it was not required before the passage of the Immigration Act of 1990 (a generally pro-immigration piece of legislation), would-be employers of one class of nonimmigrant workers, those with H-1B visas, must file a Labor Condition Application (LCA) with DOL and with the petition filed with INS. DOL does not usually act on the LCA, which describes wages and working conditions. It is simply there as a matter of public record and potentially could be reviewed by a union or other employees; further, if the foreign worker, seeking a visa as a result of the INS-approved petition and the filed LCA, says that he or she will be paid less than the LCA indicates some official questions should be set in motion.[52] Despite these potential safeguards, little has been done with these documents, a subject to which we will return.

DOL's other role in the process is that of issuing labor certifications; this process dates back to the mid-1960s. If an employer wants to hire a foreign worker, the firm files a labor certification application with DOL, and then, after it is approved, a petition with INS. The application spells out the requirements of the job, and wages and working conditions. The petition in turn usually leads to a visa for the worker, and, upon his or her arrival in the U.S., permanent resident alien status.

The point of this process, as enacted by Congress, is to allow employers to bring in foreign workers only when their wages and working conditions are such that they do not adversely affect other resident workers.

The labor certification is usually for a single, permanent job, and the employer's application is usually tailored to meet the qualifications of the worker the employer wants to hire. The employer must advertise the vacancy in the press and file it with the local office of the U.S. Employment Service. Often resident workers reply to the ads, only to be rejected for the job in question (but once in a great while someone is hired for another position with the firm in question). The labor certification process does not really test the

---

of hundreds of thousands of Mexican National farmworkers.

[52] "INS Instructs on Wages for H-1B and L-1 Petitions" *Interpreter Releases*, August 23, 1993, p. 1109. *Interpreter Releases* is a Washington, D.C.-based weekly written primarily for the immigration bar.

U.S. labor market, as it is alleged to do, but it does allow a domestic employer to hire a foreign worker under certain conditions.[53]

The third agency on the list, INS, plays a much more prominent and continuing role in the life of foreign-born S/Es than either the State Department or DOL. It is an INS inspector, at the airport, who checks the student's papers both upon first arrival and on subsequent trips back to the U.S.; it is an INS district office staffer (under some circumstances for some nonimmigrants) who approves summer employment during university years or a period of paid practical training after receiving the degree; it is an INS officer who passes on applications for adjustment from one nonimmigrant class to another, or from nonimmigrant to immigrant status. Finally, it is another INS officer who interviews the individual, now an immigrant, and passes on the application for naturalization and citizenship status.[54]

We previously defined a nonimmigrant as an alien who can stay temporarily in the U.S. while engaged in a specific, authorized activity. Congress has spelled out dozens of classes and subclasses of nonimmigrants, four of which are the most likely to include foreign-born S/Es: F-1, H-1B, J-1, and L-1.

F-1 visas are for foreign students; most undergraduates and some graduate students have this visa. Holders of this visa may hold jobs in the U.S. only with special consent (due to unexpected financial problems), and spouses of F-1s may not work.

J-1 visas are for exchange visitors, a category that covers a wide range, from student or trainee through a visiting professor. A J-1 can work in the U.S. if the work fits into his or her previously approved program (and no INS permission is needed). A J-1 spouse, with INS

---

[53] An abortive effort to weaken worker protections (and streamline the process for employers) was made by DOL in 1993; the story of the proposed Labor Market Information modification of the labor certification process, and the uproar it caused in some science and engineering circles, is told in Chapter 6.

[54] And in the unusual case where the foreign-born scientist or engineer gets into major trouble, like the green card-carrying chemical engineer who designed the bomb that blew up the World Trade Center in New York, it will be an INS officer who seeks to deport him.

permission, also may work.[55]

H-1B visas are for skilled workers, such as S/Es, and the visa carrier is expected to work, temporarily, within the U.S. economy. Of these four categories, only in this one does the employer need to inform DOL of the salary paid to the worker. This is done through the submission of the previously-mentioned LCA, which can be challenged, but rarely is.

The L-1 visa, the newest of the set, is designed to allow multinational corporations to move their managerial and professional employees from jobs overseas to jobs in the U.S.

Each of these visas is set in motion by an American institution — none of these can be acquired directly by even the most talented alien.

Any INS-recognized educational institution[56] can issue an INS form I-20, which indicates the admission of a foreign student; this leads routinely to the F-1 visa. A much smaller list of organizations, program sponsors recognized by the United States Information Agency, can issue a similar piece of paper, which leads to the J-1 visa. A would-be H-1B employer must have an approved petition from INS before seeking an H-1B visa for a temporary worker. A similar arrangement is made for L-1 workers.

The Immigration Process: The Experiences of Rajeesh Metra

Let's now look at these arrangements from the point of view of an imaginary chemist from India; we will call him Rajeesh Metra. For the sake of this presentation let us assume that he passed through each of the four nonimmigrant categories (a bit unlikely) before he secured a green card. Rajeesh's story is invented, but each of the

---

[55] Under some circumstances a J-1 visa holder may be required to return to the home country for two years prior to applying for immigrant status in the U.S., but this requirement is often waived on one of several grounds. For example, the Indian Embassy routinely waives the provision for its J-1s, the U.S. Congress waived it for Chinese students in the wake of Tienanmen Square, the USC citizen spouse of a J-1 can apply for the elimination of the requirement, and so on.

[56] INS rarely de-lists an educational organization; once in a while a private-for-profit institution that is little more than a visa mill will attract INS attention and be removed from the approved list.

elements is based on fact.

Rajeesh is from an upper-middle class Indian family and always did well in school. However, he was more restless than his hometown peers and did not relish the notion of spending his undergraduate years living at home in Madras, attending a nearby university (the Indian norm). So he took the SAT, got a high score on that, and was accepted at several U.S. universities.

He accepted the acceptance of the one (let's call it Mid-States College) that gave him the best financial package; it covered some, but not all, of his expenses. That institution also sent him the I-20, which he took to the U.S. consulate in Madras along with his passport. The consular official gave him an F visa, good until he completed his course of instruction (in this case leading to a bachelor's degree in chemistry).

Rajeesh flew to JFK Airport, waited in a long line, and spent about one full minute with the INS inspector. The inspector made a quick decision about Rajeesh; his paperwork seemed to be in order; he was on a generally "clean flight" (i.e., not one routinely used by drugrunners and illegal aliens); he met the "whole man" test in that his clothes, his passport, his airline ticket, and his demeanor all fit neatly with his stated role, a freshman accepted into an above-average American college. Further, he seemed self-confident. Rajeesh was soon on his way to Mid-States.[57]

Rajeesh earned good grades at Mid-States and enjoyed his major in chemistry. In order to meet expenses, he worked 12 hours a week in the library, a part-time, on-campus job that he could hold without INS permission. Money got tighter, however; his parents, while well-to-do in India, had an income that was quite modest by U.S. standards and could not help him very much. He decided that he

---

[57] This paragraph is based on many conversations with immigration inspectors some years ago when I conducted a study of the inspection process for INS. In Rajeesh's case, the inspector, without saying anything about it, probably noted from the passport that this was Rajeesh's first visit to the States, that the airline ticket had been purchased in his stated hometown in India several weeks before the flight, and that he was booked through to the airport nearest to Mid-States College — everything that one would expect in a legitimate case. Had Rajeesh, with the same story, been travelling on the return leg of an airline ticket purchased in New York City, for example, he would have been referred to a supervisory inspector for an explanation, and his luggage searched carefully for contraband.

had to get a full-time summer job, and for this he needed both university and INS permission.

The university said yes quickly and passed a recommendation along to INS months before the start of the summer. Time passed but nothing happened to the application. The foreign student adviser investigated and found that the application had been "remoted" to the INS inspectors at the not-very-busy Cleveland International Airport; the notion was the inspectors could do paper work between flights. Good idea, but in this case Rajeesh's application had been lost. The persistent foreign student adviser managed to persuade an INS official she knew to accept, and act quickly and favorably upon, a substitute application.[58]

So Rajeesh got the summer job, and he held onto it throughout his undergraduate years.

By late in his junior year, Rajeesh had made three decisions: 1) he wanted a PhD in chemistry, and he wanted to study under a professor at Research University who was prominent in polymers; 2) he wanted to get married to another Indian, also currently in the U.S. on an F visa; and 3) he wanted to go to Research U on a J-1 visa, so his bride-to-be could work (as spouses of F-visa holders may not).

Rajeesh then did something unusual for a would-be graduate student; he took the bus to Research U, paid a call on the professor of his choice, and considerably enhanced his admissions prospects. Research U's foreign student adviser had access to a J-1 program, and after Rajeesh was accepted as a graduate student she filed an application to adjust his status from F-1 to J-1; INS said yes. Rajeesh got married and his bride became a J-2.[59]

After a successful stint at graduate school and no contact with

---

[58] "She" is used advisedly. Foreign student advisers, as a group, are mostly female, between the ages of 30 and 55, with some graduate education but not a doctorate, cosmopolitan, outgoing, and, understandably, pro-education and pro-immigrant. They are, however, largely powerless on their campuses; see Goodwin and Nacht, *op. cit.*, p. 8.

[59] INS visas, like the dramatis personae in a Shakespearean play, are arrayed in a hierarchial order; thus an ambassador is an A-1, a lesser diplomat is an A-2, and a servant of either is an A-3; a student is an F-1, and a spouse of a student is an F-2, etc.

INS for years (the J-1 program is more relaxed than the F-1 program), Rajeesh found himself gaining a PhD but losing his J-1 status. (Rajeesh and his J-2 spouse had meanwhile given birth to a USC baby, but that did not affect the parents' immigration status.)

Rajeesh could not locate a junior faculty position but did get a postdoc job at Big Eastern U, where the Chemistry Department had a research grant on polymers. Big Eastern U's foreign student adviser filled out the forms and sent them to INS; soon Big Eastern U had permission from INS for Rajeesh to be hired as a "temporary worker of distinguished merit and ability" and INS once again adjusted his visa status, this time to H-1 (a status that in more recent years has become H-1B and now requires filing a LCA with DOL).

Meanwhile, back in Madras, Rajeesh's father's health was failing, and the old gentleman apparently had only months to live. Rajeesh, being the eldest son felt obliged to return home with his wife and baby to lend moral support and help his mother with the family finances. He resigned his postdoc with Big Eastern U, but did not lose his interest in returning to the U.S.

Months passed; his father neither died nor recovered. Rajeesh felt he had to go back to work, but felt obliged to stay in Madras, at least until his younger brother secured a bachelor's degree. With an eye on the U.S. immigration system and with his new U.S. PhD in hand, Rajeesh approached the local plant manager of a multi-national that — in the U.S. — had a commercial interest in polymers. Let's call it the PolyNational Corporation.

There was no work in advanced chemistry at the Madras plant, but Rajeesh did have an American PhD and the plant manager liked the idea of having someone with such a stellar credential on his staff; besides the manager had been a classmate of Rajeesh's father years earlier. While the work he did was, Rajeesh felt, beneath his level of training, he needed professional work in Madras, and the link with PolyNational might be useful in the future.

Rajeesh spent the next two years keeping things going at his parents' home, holding down the job at the plant, and teaching a little chemistry on the side at Madras University. Then his father died, but by then his younger brother was through university and could take over the filial responsibility to the widowed mother. Rajeesh was free to return to the U.S.

Rajeesh had served at a high enough level at the PolyNational plant — and had gotten to know some of the chemists in the home office in the U.S. — so that when the time came for his return to the U.S. he was able to secure an offer of a research position with the firm in North Carolina.

Meanwhile, because PolyNational was regarded by INS as a multi-national and because Rajeesh had held a "specialized professional" position with the firm for two years, the company sought an L-1 visa for him and L-2s for his wife and their second child, who had been born in Madras. (No visa was needed for the older child, a native-born USC.) INS approved the L petitions, the American consul in Madras issued the three L-visas, and Rajeesh was on his way back to the States.

This time, however, he chose a different route. Instead of going through JFK, where the traffic is always heavy, the lines long, and the process slowed by inspectors worried about illegals and drug dealers, Rajeesh opted for the relatively quiet Washington Dulles Airport and a flight from London on British Air. The inspector took one look at Rajeesh and his family and figured he met the "whole man" criteria. Rajeesh again made it past a gatekeeper.

Shortly thereafter Rajeesh persuaded PolyNational to file a petition with INS to adjust him from nonimmigrant to immigrant status; this is particularly easy for an individual either holding an L-1 visa or capable of securing one. These types of adjustments do not require a labor certification from DOL, as do adjustments of other professional workers.

About six years after securing the green card Rajeesh, his wife, and their younger child had their final contact with INS. Rajeesh and his wife filed applications for citizenship, were interviewed by INS officers, and, some months later, in a pleasant court-room ceremony, were sworn in as U.S. citizens while their older child, already a citizen, watched. The younger child, now eight years old, was automatically granted citizenship along with the parents.

As noted earlier, Rajeesh's contacts with INS were a little more numerous than those of most foreign-born S/Es. A more typical progression would be an arrival as a graduate student (on a J-1), movement through a temporary U.S. job, (as a postdoc, perhaps, on an H-1B), and then on to immigrant and naturalized citizen status. Rajeesh had relatively good luck with INS, except for the lost application to work one summer, but this, as we show subsequently, is a reflection of a bigger picture — that while the immigration process complicates and sometimes slows the life of the foreign-born S/Es,

The Immigration Process: Numbers and Decisions

it rarely denies them a desired benefit.

Although Rajeesh managed to use four different nonimmigrant classes, these classes are not used equally by all foreign-born S/Es. As Exhibit Nine shows, scientists are more likely to use J-1 visas, while engineers are more likely than engineers to use the H-1B and L-1 routes. The fairly new Canadian Free Trade Agreement offers a rapidly growing path into U.S. professional jobs, trod by native Canadians and non-Canadian-born persons who have secured Canadian citizenship.[60]

How plausible is our story of Rajeesh's many successful contacts with the U.S. immigration system? And if plausible, and if the numbers of admissions of foreign-born S/Es continues to rise, why is there so much tension between INS on one hand and America's universities on the other? The tension was clearly visible at the previously-mentioned NAFSA convention in Chicago.

What is not generally recognized, and certainly not discussed by either INS or its pro-immigration critics, is that the U.S. immigration system generally says "yes" to alien applicants, no matter what the question (unless it involves illegal entrants encountered by the Border Patrol). For example, the following approval percentages are typical,

---

[60] The usual waiting time for citizenship in Canada is three years, as opposed to five years in the U.S.; though most immigrants to Canada stay there, some use Canada as a stopping place en route to the U.S. The CFTA, unlike the arrangements in the European Community, permits the easy cross-border movement of only professional workers, not all workers as in the EC.

**Exhibit Nine: Nonimmigrant Admissions
of Scientists and Engineers,
FY'93, by Visa Class**

(estimates based on INS data)

	H-1B	H-2B and H-3	J-1	L-1	CFTA	Total
Occupation						
Engineers	17,766	987	2,804	15,349	2,568	39,474
Computer and Mathematical Scientists	7,757	39	284	1,843	923	10,846
Natural Scientists	3,841	16	3,829	904	754	9,344
Total	29,364	1,042	6,917	18,096	4,242	59,664

Visa classes: H-1B, aliens in specialty occupations holding temporary jobs; H-2B, non-agricultural temporary workers (a class rarely used by professionals); H-3, industrial trainees, another seldom used class; J-1, exchange visitors; L-1, multi-national employees; and CFTA, Canadian Free Trade Agreement.

For source and estimation techniques, see Exhibit One and the Appendix.

---

even though they deal with the actions of three different agencies:

Benefit Sought, Agency Concerned	**% Approved**
Nonimmigrant visas, State	88.7%
Immigrant visas, State	92.3%
Sample of H & L petitions filed with INS, Atlanta, Ga.	93.0%
Farmworker (SAW) legalization applications, of those decided, INS	93.5%
Regular legalization applications, of those decided, INS	94.5%
Naturalization applications, INS	98.0%
Labor certifications, science and engineering, DOL	98.9%

Data are global for the State Department, nationwide for DOL, and for all INS benefits except the five-month sample in Atlanta. Immigration managers do not seem to be very interested in approval/denial rates *per se*; these rates had to be calculated by adding denials to approvals, and then calculating the percentage.[61]

The first percentage above (88.7%) sets the tone. More than nine million people came in off the streets of London, Rome, Mexico City, and scores of other places around the globe seeking to enter the U.S. as tourists or in other nonimmigrant roles. Few of these applications were supported by U.S. institutions, fewer still were prepared by professionals, or were filed on behalf of people with professional occupations, but close to 90% of them were approved.

Given that sort of approval rating for run-of-the-mill nonimmigrant applications, it is no surprise that so a high percentage of the labor certifications were given the green light. *All* were filed by a U.S. institution, usually a corporation or university; in all likelihood most of the applications were prepared by experts (lawyers or foreign student advisers), and all related to people with extensive

---

[61] The visa rates were calculated from the State Department's *Report of the Visa Office: 1988*, tables I and XX, using "ineligibility findings" minus "ineligibility overcome" to arrive at the total number of denials; the Atlanta data were from D.S. North and Paoze Thao, "Layer Cake: The Foreign-Born and Atlanta's Labor Market," a report to the National Commission on Employment and Training, Washington, D.C., 1986, p. 53, and secured from a review of the files of the INS district office; legalization data are from INS' *Immigration Reform and Control Act: Report on the Legalized Alien Population*, Washington, D.C., 1992, table 1; naturalization data are from *INS 1991 Statistical Yearbook of the Immigration and Naturalization Service*, Table 41, with petitions denied in 1991 being compared to persons naturalized that year; and the labor certification data for FY 1992 originated in the Employment and Training Administration of the U.S. Department of Labor and were published in *Chemical and Engineering News*, April 26, 1993, p. 6.

There is another technique for measuring approval rates, rather than comparing the approvals and denials of cases reaching a final decision (as we have here); one can compare the total intake of applications with the total number of approvals. This system lumps together denials with applications that have been abandoned for a variety of reasons (death of applicant, corporation going out of business, decision not to provide requested additional information to the agency, etc.). Using this approach, the ratio of approvals to denied and abandoned cases has been about 80% for DOL-handled labor certifications in recent years, and about 65% in FY'93; these percentages cover all applications, including those for low skill jobs, where the abandonment percentage is higher than average. Interview with Flora Richardson, director of DOL's labor certification program, Dec. 17, 1993.

science or engineering credentials.

In short, the more skilled the population reaching out to INS, the more likely a favorable response, but the agency usually says yes to anyone who asks. It was probably appropriate that INS always said yes to Rajeesh, but it was also totally predictable. It is also true that if the U.S. immigration process were staffed more generously there might be a higher percentage of denials in every category listed above.

As an old hand in the immigration business, the late Lauren Lawrence once explained to the author: "It takes much, much more time to say 'no' to an applicant than to say 'yes'; if you approve the application you never see the applicant again and never have any more work to do on the case; if you say 'no' there are usually appeals, letters from Congress, and telephone calls."[62] The implication was that an overworked bureaucrat will think twice before rejecting an application.

We conclude this part of the discussion with a vignette from the 1992 NAFSA session. Several hundred foreign student advisers (all of whom paid an extra fee for the session) gathered in a hotel ballroom to discuss the latest governmental rulings on foreign students; they listened and talked for hours about special rules for Chinese students, the then-existing confusion about summer jobs for F-1s[63], the wages that had to be paid to postdocs, and other detailed issues. The INS staff present were stiff, young, and uncomfortable. Then the leading speaker asked a question; he was Stephen Yale-Loehr, a prominent immigration lawyer and then co-editor of *Interpreter Releases*, the authoritative newsletter for the immigration bar. Pro-immigrant and pro-education and in tune with his audience,

---

[62] Lawrence served, *inter alia*, as consul-general in Manilla, which has an enormous amount of visa activity, and as Director of the Passport Office.

[63] Throughout the spring and early summer of 1992, INS was wrestling with what it saw as new restrictions in the 1990 Immigration Act regarding summer employment for F-1 students; it apparently felt that a remarkable provision, generally known as the McDonald's Amendment, which allowed fast food chains special access to F-1 students, established the only kind of summer employment for such students. The universities felt that this was an *additional* program, not a *substitute* for long-established earlier programs, and eventually convinced INS that this was the case. But in the interim, off-campus student employment for F-1s was lost for the summer. For more on this flap see *Interpreter Releases*, July 20, 1992, pp. 861-862.

he asked: "In the final analysis have any of your institutions ever lost a case to INS? If so raise your hand."

Not a single hand was raised.

Why then the tension in that room, and elsewhere, between INS and the universities? Part of it is cultural, the well-educated staff of educational institutions dealing with what they might view as a blue-collar police agency (the career leadership of INS, until recently, consisted largely of ex-Border Patrolmen); part of it is ideological, with the universities confident that they have a better grasp of appropriate public policy than any federal agency; part of it is frankly ineptitude on the part of some INS staff and its then leadership; and part of it related to the amount of time and energy the foreign student advisers had to use to determine the exact status of the rules, and then to comply with them.

In short, while the government rarely says "no" about a foreign-born S/E or any other foreign student, it does not always say "yes" clearly and quickly; sometimes it mumbles. Further, it is not always consistent, and it often creates complications and extra work for both the foreign student advisers and their advisees. Finally, sometimes the government presses policies that the universities dislike; for example, trying to set some standards for postdocs wages or seeking to protect U.S. workers from foreign student competition.

Thus the tension in the room.

Chapter Four

The Roles of Foreign-Born Scientists and Engineers in the U.S.

Overview

What do the foreign-born scientists and engineers do in the U.S.?

First, most of those who secure graduate degrees in the U.S. tend to stay here after graduate school, rather than return to their home countries.

Second, while in graduate school, they generally perform well and apparently are better funded than USCs.

Third, two quite different patterns in industry exist for foreign-born S/Es. There is the older, larger, slowly growing, less controversial pattern of foreign-born S/Es moving, one at a time, through nonimmigrant and immigrant status before becoming naturalized citizens, often following graduate work in the U.S. Then there is the newer, currently smaller, fast-growing, and very controversial utilization of foreign-trained S/Es, particularly in the computer industry; these workers are brought to this country in groups as nonimmigrants at what many believe to be exploitive wages.

This chapter deals with the last two points in some detail, but first we look briefly at the first point, the matter of return migration, on which we have the least information.

Very little is known about the long-term location of one of subpopulations of interest, the S/Es who come to the U.S. as mature professionals, those immigrants who come directly to work, rather than to attend graduate schools. (As noted earlier there appears to be about one overseas-trained foreign-born S/E for every four foreign-born S/Es with some U.S. training.) By definition, the smaller of these immigrant populations probably reaches our shores with an average age in the late twenties or early thirties. As Ravenstein pointed out so many years ago, all migration streams create eddies, and there is always some return migration,[64] but the extent to which mature foreign-born S/Es return home after settling for a while in the U.S. is not known. It is safe to assume, however, that neither the numbers nor the percentages are large.

More information exists on the ultimate location of those foreign born who have secured graduate degrees, particularly PhDs in the U.S.: as a rule of thumb it looks as if at least 60% stay here after receiving the doctorate.

Finn, of Oak Ridge Institute for Science and Education, who has followed these populations for years, conducted a study of the foreign born securing BS and MS science and engineering degrees during the period 1976-1979, and found that 42% of the bachelor's holders and 62% of the master's holders were still in the U.S. in 1982. Similarly, of the 1980-1981 foreign-born doctorates, 62% were also found to be in the U.S. in 1982.[65]

Using another data source, special tabulations from nonresponses to the National Research Council's 1981 and 1985 Surveys of Doctorate Recipients, Finn estimated the *emigration* rate of new foreign-born science and engineering PhDs as follows:[66]

Naturalized USCs	1.3%
PRAs	13.8%
Nonimmigrants	45.0%

---

[64] Ernest George Ravenstein, *Laws of Migration*, Arno Press, New York, 1976, reprint of original, in two parts, in Vol. 48, Part 2, June 1885, and Vol. 52, Part 2, June 1889, *Journal of the Royal Statistical Society*, London.

[65] See his NSF report *Foreign National Scientists and Engineers, op. cit.*; these data can also be found in: National Research Council *Foreign and Foreign-Born Engineers in the U.S.: Infusing Talent, Raising Issues*, Washington, D.C., National Academy Press, 1988, p. 66.

[66] See *Foreign and Foreign-Born Engineers in the U.S., op. cit.*, p. 97.

Since nonimmigrants are much more numerous among the new PhDs than the other two civil status classes, this works out to a retention rate of about 60%.

The 1989 *Summary Report*, which had a special section on foreign-born PhDs, reported increasing rates of U.S. retention of this population. "In 1973, 51% (or 1,595) of the foreign PhDs with definite postgraduation plans planned to stay in the United States after graduation; by 1989, the proportion had grown to 63% (of 2,904 PhDs)."[67] So a larger percentage of a growing population of foreign-born science and engineering PhDs seems to be staying in the U.S.

The Foreign Born in Academe

The Graduate School Experience

It is difficult to generalize on the graduate school experience of foreign-born PhD candidates. In the preface to their book, Bowen and Rudenstine commented about, "how hard it has been to obtain answers to even the most elementary questions concerning graduate education....Another relevant consideration is the simple fact that graduate education is departmentally-based. The decentralization of activities and responsibilities...complicates enormously the task of even describing the process in anything like general terms."[68] If the former president of Princeton and the then-future president of Harvard have these problems with this subject, think of the challenge to the ordinary mortal! Nevertheless, the following sketch may be helpful, particularly for those who have not gone through this experience in this country.

Graduate education in science and engineering is not only departmentally-based, it is often tailored to the individual student; it is also narrowly focused intellectually. Graduate education can be divided into two segments, the preliminary work leading up to the nearly universal qualifying examinations (quals), and then the dissertation research.

---

[67] *Summary Report 1989, op. cit.*, p. 46.

[68] William G. Bowen and Neil L. Rudenstine, *In Pursuit of the PhD*, Princeton University Press, Princeton, N.J., 1992, p. xv.

In some departments all graduate students must take at least a specified, core set of courses in preparation for the quals. In others, the courses are offered but students are not mandated to take them. The quals are given periodically, and typically the student takes the examination when the student feels ready. Students failing the quals are sometimes given the opportunity to pursue a master's degree as a consolation prize.

With quals completed and other requirements met, the student is then expected to conduct an original piece of research that will lead to a dissertation. This is to be a substantial body of work, and the research usually takes several years. Choosing the topic, perfecting a proposal, and finding a faculty member to supervise the project may be a year-long process in and of itself.

Often the subject chosen will need to fit into two different systems: the student's area of interest and academic background; *and* the interests of the external funding source that is providing the research assistantship that funds the graduate education. This presumably can be a troublesome fit under some circumstances, but it is a subject we did not research. Sometimes the stated interests of the external funding source are broad enough so that any plausible area of study will be acceptable.

The Bubbles Project

Speaking of "any plausible area of graduate study," we found that the U.S. Navy's Office of Research was supporting some work at the Department of Oceanography at Dalhousie University in Halifax, N.S.; curious as to what the Pentagon was buying from a non-U.S. institution, we asked about the nature of the research, and we were told "a study of bubbles in seawater."

"Why is the Navy funding this?" we asked.

"I guess they just want to be in touch with research in the field."

The work at Dalhousie may be significant and perhaps even of some potential military application, but Dalhousie should be grateful that former Senator William Proxmire is no longer in a position to make his Golden Fleece Awards.

If the student's research goes well, the results are described in a dissertation, which is presented usually in both written form and orally to a review committee of professors. If this dissertation and its defense are also successful the long process is over, and the PhD is awarded.

A graduate education at an American university is less of a socializing experience for a foreigner than a typical undergraduate education. Undergraduates, at residential institutions, often live together in dormitories, or sometimes fraternities or sororities. They take a broad range of classes with one another; they often eat in institutional settings; they can engage in sports or other structured extra-curricular activities. The academic work is only as hard as the student wants to make it. In short, even the shyest undergraduate is forced into constant contact with a cross-section of other students.

At the graduate level classes are few and small, or non-existent. One lives in an apartment, and there is no communal dining. Structured extra-curricular activities are rarely part of the scene (though one can invent one's own non-academic life). The student focuses on a small piece of knowledge, sees only other graduate students in the same discipline and a few professors, and works very hard. In short, one may see a much narrower part of America as a graduate student than as an undergraduate.

There is another interesting comparison, this one within grad schools. Nerad and Cerny suggest that there are a series of sound reasons why science and engineering grad students are more likely to finish the degree, and to do so more quickly, than those in the non-sciences: in science and engineering, the graduate student's research is often of real significance to the supervising professor, whereas this is usually not true in the other fields; there is much greater funding of science and engineering graduate students than others; and science and engineering usually lack additional requirements (such as languages) that nonscience programs tend to have.[69] The foreign born in American grad schools seem to know this; they certainly concentrate in S/E.

Foreign-Born and Native-Born Graduate Students: A Comparison.
Although the data are far from complete, it appears that foreign-born S/Es, at least at the PhD level, do very well, probably better than their native-born peers, at U.S. graduate schools.

---

[69] M. Nerad and J. Cerny, "From Facts to Action: Expanding the Educational Role of the Graduate Division," *Communicator*, Council of Graduate Schools, Washington, D.C., 1991.

One hears abundant comments from science and engineering faculty members that the foreign-born students come to graduate school with good preparation and good work habits. Many have done some graduate work in their home countries before they arrive in the U.S., as well.[70]

But they do bring some handicaps in addition to any problems they have with the English language. We explored this in some depth with Kevin Aylesworth, then a fairly new PhD in Physics, and founder of the Young Scientists' Network.[71]

"The Asian PhD students have an excellent background in theory and work very hard," Dr. Aylesworth said. He recalled his first year in graduate school (at the University of Nebraska) as a "nightmare" as he struggled to keep up with the better-prepared foreign-born graduate students.

But the Americans do much better, at least initially, in the laboratory, he said. There were two reasons: First, there is a tradition among the Asian elites for the males not to work with their hands; that is something for servants to do. For similar reasons, elite male students are not taught shop in primary or middle school, so the arriving PhD candidates have no background even in such household crafts as carpentry. Second, most of the foreign-born have not lived around as much machinery as young Americans and are initially awkward using lab equipment, fearing that they will break something. The American male, who may have been fixing bicycles and automobiles all of his life, is more relaxed with the same equipment.

It takes a year or so for the Asians to become familiar with lab equipment and for the Americans to change their study habits, and after that things are pretty equal — that's Dr. Aylesworth's cheerful

---

[70] One source said that most Mexican Nationals coming to the U.S. to do graduate work in science or engineering have already served on the faculty of their universities.

[71] A long interview and several telephone conversations with Dr. Aylesworth in April 1992. Dr. Aylesworth was one of the witnesses at the April 8, 1992, hearing on the National Science Foundation's study predicting shortages of S/Es. That hearing, before the House of Representatives' Subcommittee on Investigation and Oversight of the Government Operations Committee, is described in Chapter 6. Dr. Aylesworth was subsequently elected as a petition (i.e., non-Establishment) candidate to the Council of the American Physical Society.

assessment.

Turning from the qualitative to the quantitative, there are two measures of any population's experience in graduate school: how long did it take to obtain the degree? and what percentage of those admitted secured degrees? There is more data on the first measure than the second; the number of years between admission to the graduate school and the receipt of a degree (for those who receive the degree) is known as Elapsed Time to Degree (ETD) or Registered Time to Degree (RTD).

Although Bowen and Rudenstine do not deal with citizen/noncitizen patterns in graduate school, they do provide extensive information on the ETDs of various groups of students receiving doctorates. The EDT is longer, on average, in the humanities (8.61 years) than in the physical sciences (5.89 years), with engineering at 6.22 and life sciences at 6.34. Further, the ETD for women is generally longer than it is for men.[72]

With the new foreign-born PhDs being mostly in the science and engineering fields and with the foreign born having a higher percentage of males than new PhDs generally, one might suspect that they complete graduate school more quickly than native-born PhD candidates, generally; and that is the case.

But when we examine groups of USCs and nonimmigrants in the same science and engineering fields, as *Summary Report 1991* does, we find that there are few differences. The years registered in graduate school range from 6.1 to 6.7 for the USCs, and from 6.1 to 6.5 for the nonimmigrants.[73]

Attrition rates are harder to examine than years in graduate school because while all the newly-minted PhDs can be asked about their time in graduate school, it is harder to track down everyone who enrolled in graduate school twelve to twenty years earlier to find out what happened to them. Some may have left the first university and secured a degree elsewhere, for example.

It is the impression of two scholars who have spent much time on

---

[72] Bowen and Rudenstine, *op. cit.*, pp. 130-132. These authors use the ETD concept and the *Summary Reports* uses RTD; the definitions appears to be similar, if not the same.

[73] *op. cit.*, p. 16.

graduate school education issues that the foreign born are less likely to drop out than the native born. As Ronald G. Ehrenberg of Cornell sees it, the PhD candidate from Indiana now enrolled in Michigan has many more options than the candidate from India. The Hoosier, without losing face, can decide to become a lawyer or go to work in industry; the Indian, however, faces strong community and family expectations that he will return with the PhD.[74]

Similarly, Maurice Yeates, Executive Director of the Ontario Council on Graduate Studies, told me that in Canada the foreign-born graduate students were more likely to graduate and likely to do so more quickly than their Canadian contemporaries. There was also a higher percentage of males among them than among the Canadian students.[75] There is little statistical confirmation of what appears to be sound judgements on this matter; the forest of statistics generated by the government on foreign-born S/Es does not seem to touch on this subject directly, but there does seem to be some indirect evidence, shown in Exhibit Ten. When one compares the percentage of foreign born among graduate students in science and engineering (using two different 1991 measures) with the percentage of foreign born among those with new doctorates, one finds the latter percentage is always higher. (This was the case in earlier years as well.)

Since we know that foreign-born and native-born S/Es take about the same time to reach the PhD in the U.S., the difference can be explained in only two ways: the foreign born are less likely to drop out of PhD programs than the native born, and/or a higher proportion of the foreign born seek doctorates (rather than the master's degree) than do the native born. In any event Exhibit Ten suggests that the foreign born are doing a little better in graduate school than their native-born peers.

---

[74] Conversation with Professor Ehrenberg, Director, Institute for Labor Market Studies, Cornell University, at an NRC seminar, September 25, 1992. He also noted that in the past there was another factor at work: during the Vietnam War the draft kept many American men from dropping out of grad school.

[75] Personal communication from Dr. Yeates; see also his "Doctoral Graduation Rates in Ontario Universities: A Discussion Paper," Ontario Council on Graduate Studies, Toronto, Ont., 1991.

Exhibit Ten: Some Indirect Measures of Foreign-born Students' Prowess in Graduate School

A Comparison of 1991 Enrollment and Doctorate Awards Data
(percent of foreign-born in each category)

Discipline	Graduate Students in Doctorate Granting-Institutions	Full-time Graduate Students in all Institutions	Doctorates Granted
Physical Sc.	36.9%	39.5%	40.3%
Earth Sc.	20.4%	24.0%	26.8%
Mathematics	35.2%	41.8%	54.9%
Computer Sc.	33.1%	47.1%	51.3%
Biology	24.2%	26.0%	27.5%

Source: All data from J.G. Huckenpöhler, *Foreign Participation in U.S. Academic Science and Engineering: 1991*, NSF 93-302, Washington, D.C., 1993. The second column is from Table A-10, the third is from Table A-12, and the fourth is from Table A-17.

Funding the U.S. Education of Foreign-born Scientists and Engineers

This section describes the financing of the U.S. graduate education of foreign-born S/Es: the sources of the funding (U.S. and overseas), the ways that universities think about financial assistance for different groups of students, and the mechanics of funding the foreign-born graduate students. As noted earlier, while some of the foreign-born S/Es come to America with their education completed, most seek one or more degrees in the U.S.

U.S. or Foreign Funds? Who pays for the American education of the foreign-born S/Es? This is not an idle question. If the migrants (or other foreign sources) pay for it, then we are dealing with funds beyond America's control. If U.S. resources are used, on the other hand, some of them could be reprogrammed to support education for native-born students, or for other purposes.

One encounters two diverse flows of information on this basic point. First, there are the box-car numbers favored by the media, such as the notion that the "export" of education from the United States produced, according to a U.S. Commerce Department estimate, a balance-of-payments surplus of $5 billion a year in 1990. Another often-cited estimate, this from the Institute of International Education (IIE) is a little more conservative. The statement is that by a wide margin the principal source of funding for all foreign students is their own, and their families' money, suggesting a large influx of funds. These concepts are sometimes mentioned against the backdrop of total balance of payments data, which, in recent years, have always been heavily in the red.

The other flow of information, particularly regarding the graduate educations of foreign-born S/Es, runs almost precisely in the other direction. Every foreign-born scientist or engineer the author met told me that the funding of his or her entire graduate education came from U.S. funds;[76] deans and financial assistance officers we interviewed at the research universities say that with a few exceptions, this is the case. Further, the oft-cited *Summary Reports* of the National Research Council indicate that the foreign born receiving doctorates identified American institutions as the principal source of their funding in 72% to 89% of the cases.

Why this apparent discrepancy? There are several reasons. First, the $5 billion Commerce Department estimate is not a Commerce Department estimate at all; it is a single journalist's misreading of a single Commerce Department statistic. The writer, unfortunately, was working for the widely read *Washington Post*, and the erroneous number has since developed a life of its own, appearing in many conversations with foreign student advisers, in *Business Week* and in the *New York Times*. It will probably be with us for years to come.[77]

---

[76] "Why did you choose that institution for your graduate work?" I asked one foreign-born physicist. "They made the best offer; they were the only school that would give me full funding from the very first term of graduate study," he replied.

[77] Paul Blustein "A Hidden U.S. Export: Higher Education: American University Degree is a Highly Prized Commodity Among Foreign Students. Especially Asians," *Washington Post*, February 16, 1992, section H, Financial, p 1. See also a letter to the editor of the *New York Times* from Harvard University's Dorothy S. Zinberg on May

What the *Post* reporter apparently did was to misunderstand a single number in the annual report on balance of payments for 1990 that appeared in *Survey of Current Business*, the Commerce Department publication. He found (in Table 3) that the U.S. exported $4,828,000,000 worth of education, (i.e., that absorbed by foreign students in the U.S.)[78]. The reporter missed — partially because Commerce does not publish data on this point — an important variable: the source of funding for this education. The $4.8 billion is a gross figure, the value of education received in the U.S., with no data provided on who paid for it.[79] Commerce does collect data on the source of the funding, but does not release them separately (these are small potatoes in the grand sweep of international trade). So there is no Commerce Department data on this point beyond an estimate of the value of the education received by foreigners in the U.S. and by U.S. students abroad.

A second reason for the seeming discrepancy on this point — who pays the bills — relates to how the Institute International Education (IIE) collects data on the subject, and how others interpret it. The most memorable number and concept is IIE's statement that the

---

12, 1992. She wrote: "Last year [presumably 1991] the education of foreign students netted the United States some $4 billion, according to the Department of Commerce. These profits should somehow be invested in the education of our citizens." We do not have a citation for the *Business Week* use of the surplus estimate, but it apparently appeared in 1992.

[78] Christopher L. Bach, "U.S. International Transactions, Fourth Quarter and Year 1990," *Survey of Current Business*, March 1991, U.S. Department of Commerce, Washington, D.C., pp. 34-68.

[79] In Table 1 of the previously cited article, payments for the education of foreign-born students are lumped with other transactions on lines 23 (student earnings), 30 (U.S. Government grants), 32 (fellowships from non-government sources), and 47 (a miscellaneous category including payments from abroad by individuals). A further potential source of confusion in these articles, which are written annually, is the use of the term "services and investment income, net" in connection with education (see Table L in these articles). The net figure is for the gross value of education provided in the U.S. to the foreign born, minus the gross value of education provided overseas to Americans. Neither of these totals speaks to the *source* of funding for the education provided. The author is grateful to William McCormick of the Balance of Payments Division, Bureau of Economic Analysis, U.S. Department of Commerce, for his explanation of this complex accounting system.

primary source of funding for foreign students in 66.2% of the cases is "personal and family funds."[80] The second source of funding in recent years has been "US college/University" with the percentage being 19.0% in the most recent survey; no other primary source exceeds 6%.

Two observations can be made about IIE's data. First, funding patterns are much different for undergraduate and graduate students; deeper within *Open Doors 1991-1992* one learns that personal and family funds are the primary source for 81.7% of the undergraduate students and 46.2% of the graduate students.[81]

Second, the IIE methodology is open to question. One would wish that a different question had been asked of a different population. Ideally, a sample of foreign students would be approached directly, and asked: "What percentage of your funding came from which sources?"

Instead, IIE asks the universe of college officials responding to the annual survey to provide estimated percentages of the primary source of funding of the foreign students at their institutions. The questionnaire's instructions, which IIE, to its credit, publishes each year, provide no help to the respondent as to how to gather the data sought.

Over and beyond the data collection problem, there is the nature of the question itself. Any such question about primacy, as opposed to distribution, can produce misleading implications. For example, each year IIE reports that current off-campus student employment is the primary funding source in 2.3% of the cases. On its face this may be totally accurate, but the question fails to cope with the reality of a lot of part-time off-campus employment by students. Perhaps 2.3% secure more funding from this source than any other, but, on average, perhaps 10% of students' funding comes from off-campus jobs.

There is also a lesser problem — one that could be addressed at no expense to IIE. The survey form, which IIE prints each year, arrays the possible primary sources of funding in the sequence shown below, with the replies, for 1991-1992 also displayed.

---

[80] *Open Doors 1991-1992*, Institute of International Education, New York, N.Y., p. v.

[81] *ibid*, Table 9.4.

Primary source of funding (col. 1)	1991-92 estimates (col. 2)
Personal and family	66.2
US College/University (grants, TAs, RAs etc.)	19.0
Home government/university	5.2
Private (US) sponsor	2.8
Private (foreign) sponsor	2.0
International Organization	0.5
Current Employment	2.3
Other (please specify on page 4)	0.6

Source: column 1 from *Open Doors: 1991-1992*, p. 149; column 2 from p. v.

Note the neat fit between the distribution of answers and the array of choices offered. It is only at the 2.3% level that a response gets out of line. This may not be intentional but it certainly is avoidable.

Politicians know that ballot order, particularly when a number of people are seeking the same office in a primary, is so important that state laws are passed to specify how drawings must be done for ballot position. Political pollsters wanting to be careful and fair in an oral survey often give the respondent an open-ended question, rather than a series of set choices, to avoid suggesting a particular response.[82] One wonders what would happen to the IIE's estimates if they scrambled the sequence of suggested answers in a future survey.

One final point on the IIE survey: the response rate. Some of the problems noted above and the sheer labor of trying to answer the question accurately, have taken their toll. Each year *Open Doors* contains a statement like the one in the 1991-1992 report: "the response rate to this particular question is the lowest of any asked on the *Open Doors* survey. Information on source of funds exists for only 53% of the students... which may or may not be representative

---

[82] Some months after this sentence was written the following headline was printed in *Science Times*: "Pollsters Enlist Psychologists In Quest for Unbiased Results." *New York Times*, September 7, 1993, pp. C1 and C11. The article was about efforts to form questions carefully to avoid leading the respondents. We had no response to a long letter we wrote to IIE on the subject.

of the entire foreign student population."[83] These caveats do not, however, prevent people in academia from saying that "education is a $5 billion plus in the balance of payments" and "two-thirds of the moneys for foreign students come from their families abroad."

IIE is a large organization and, as non-profits go, a relatively affluent one. One wonders why it continues to produce statements on the source of funding of the education of the foreign born when it appears to be operating from such a shaky data base; maybe it has no particular motivation to secure better information on this subject.

Now let us turn to another source of data, dealing directly with the leading experts on who pays for the graduate education of scientists and engineers, the students themselves. This is the previously cited annual survey of America's new doctorates conducted by the National Research Council. Coverage is comprehensive because if you do not fill out the form most institutions will not award you your degree. The new noncitizen PhDs reported the following in 1991: [84]

Discipline	Primary Source of Funding[85]			
	Personal	University	Federal	Other
Phys. Sci. and Math	5.2	88.4	0.5	5.9
Eng.	9.9	78.6	0.7	10.8
Life Sci.	9.2	68.9	3.5	18.4

Thus, a huge majority of the new, noncitizen PhD scientists and

---

[83] *Open Doors 1991-1992*, pp. 36-37.

[84] These calculations are based on *Summary Report 1991*, *op. cit.*, table 8. This survey uses these categories for the new PhDs: U.S. citizen, permanent visa holder, and temporary visa holder; it does not gather information by place of birth. We have merged the two alien categories in this and some other tabulations.

[85] The relatively high percentage for other sources for funding PhD programs in the Life Sciences reflects, among other things, the greater extent to which foreign governments support studies in this field, particularly in agriculture, than they do in others, and the massive support provided by the Howard Hughes Medical Institute.

engineers reported that American sources provided their primary source of funding. (Both NRC and IIE note that students, even PhDs, have some trouble with the university/government distinction; often funding that starts with the federal government, and flows through the universities, is identified with the latter by the students.)

What is equally interesting, and never mentioned in academe, is that *the further you are from U.S. citizenship the more likely you are to secure American funding* for your graduate studies in science and engineering. This pattern has prevailed for years. For example, regarding 1991 recipients of doctorates in engineering one sees:

Civil Status	Primary Source of Funding (Engineer Doctorates '91)	
	Personal	**Univ. & Federal**
USCs	21.4	64.8
PRAs	18.1	76.4
Nonimmigrants	8.6	79.7

Source: *Summary Report 1991*, table 8.

So a large majority of all PhD candidates in engineering are funded by U.S. sources, but the majority is largest among those graduate students holding temporary visas.

Another, albeit indirect, measure of the extent to which foreign-born S/Es are well treated and well funded by U.S. educational institutions is the extent to which the various citizen/alien categories find themselves in debt upon graduation. The percentage of the holders of new doctorates in 1991 reporting *no* debt were as follows:

Civil Status	Discipline		
	Phys.Sc.	**Eng'g**	**Life Sc.**
USCs	39.7	49.0	37.1
Green Cards	65.7	65.0	55.1
Nonimmigrants	78.9	70.8	70.1

Debt data by civil status and discipline is displayed in Exhibit Eleven for those receiving U.S. science and engineering doctorates in 1991. It shows that nonimmigrants were *less than half as likely* to incur debt as citizens, and that Blacks in science and engineering carried the largest cumulative debt at graduation of any group.[86]

To summarize, the large majority of science and engineering graduate students are supported by U.S. sources, with the noncitizens being treated more generously than citizens in virtually every category. While this may be appropriate public policy (the foreign-born graduate students come from nations with lower income levels than our own), there can be no argument that the foreign-born graduate students in science and engineering secure their *graduate educations* largely at American expense.

But how about the total costs of their education, from birth to PhD? It is often pointed out that, assuming that the foreign-born scientist or engineer stays in the U.S., the U.S. benefits from the costs of educating that person up to the point when he or she leaves his or her homeland, usually at the bachelor's level. What is less often mentioned is the relative cost of K-16 education vs. graduate education; the former is characterized by large classes and minimal equipment, whereas graduate education is more likely to involve very small classes, one-on-one interactions with a professor, *and* very expensive equipment.

A recent National Research Council report touched on this point, saying: "Using U.S. costs, the total investment in a B.S. degree from birth is probably about one-third of the societal cost for a supported (through a teaching or research assistantship) PhD graduate."[87]

Assuming the normal pattern of K-16 education overseas and graduate education in the U.S., we have a situation in which three-quarters of the total educational costs are borne in the U.S. and only one quarter in the home country.

---

[86] It is sometimes suggested that Black graduate students are more likely to send money to their parents than other USC and foreign-born graduate students, thus increasing the likelihood of debt in graduate school.

[87] National Research Council, *Foreign and Foreign-Born Engineering in the United States: Infusing Talent, Raising Issues*, National Academy Press, 1988, p. 19.

Exhibit Eleven: Debt Status of New PhDs in Science and Engineering, 1991

(Percentage with Debts and Mean Debt for Debtors)

New Doctorates	Life Science	Physical Science	Engineering	Mean
Black				
with debt	60.2%	55.0%	60.5%	59.0%
mean debt	$10,700	$9,318	$10,192	$10,255
US Citizen				
with debt	62.9%	60.3%	51.0%	59.7%
mean debt	$8,887	$8,569	$8,302	$8,652
Immigrant (PRA)				
with debt	44.9%	34.3%	35.0%	37.9%
mean debt	$10,205	$8,681	$9,162	$9,421
Temp. Visa				
with debt	29.9%	21.1%	29.2%	26.6%
mean debt	$10,177	$8,084	$10,189	$9,348

Source: unpublished data from the National Research Council's on-going *Survey of Earned Doctorates*. The full citation is: "Survey of Earned Doctorates, sponsored by five Federal Agencies: National Science Foundation (NSF); National Institutes of Health (NIH); U.S. Department of Education (USED); National Endowment for the Humanities (NEH); and the U.S. Department of Agriculture (USDA); and conducted by the National Research Council (NRC)."

Note: Average debt levels were estimated by the author. NRC provides four ranges of indebtedness: under $5,000, $5,000-$10,000, $10,000-$15,000, and more than $15,000. For averaging purposes $2,500, $7,500, $12,500, and $17,500 were used for the four ranges. The total doctorates in each of the four categories were: Black, 166; USC, 9,950; PRA, 1,015; and temporary visa, 6,149.

Thus even though the, say, 28-year-old Indian with a new PhD in chemistry has been in the U.S. for less than one quarter of his life, much of the cost of his lifetime education has been funded by American sources.

An Examination of the Economics of Different Kinds of University Financial Assistance. If one looks at the flow of funds from the point of view of a student (or a parent), all student assistance is welcome, and it all helps pay the bills. People speaking for higher education often tend to use the same approach — all assistance to all students is designed to make the university a better place, and the university tries to distribute these benefits equitably.

On closer analysis, one finds that the university is a series of markets, somewhat connected with one another, that have many of the economic characteristics of the outer world. Different kinds of students play different roles within the institution. Let us think about four categories of them.

Class A. The student with full funding from external sources, in any discipline, at any level, helps the university stay afloat, even though that student is, to some extent, subsidized by an endowment or public financing of the institution. Class A students pay more than their share of student-borne costs. They play no role within the university's own labor market.

Class B. The scholarship student, either as an undergraduate or as a graduate student who does not work as a TA or RA, is a drain on the system. Perhaps the student is very welcome, has an excellent mind, is a valued athlete, a member of an under-represented minority, or all of the above, but that student is a financial burden, nevertheless. Class B students pay less than their share of student-borne costs. Some of them serve as part-time, low-skilled workers in the university labor market, working as waiters, library aides, etc.

Class C. The graduate student who secures some of his financial support in return for work as a TA simultaneously receives money from the university while providing needed para- or junior professional labor in the university's labor market. As workers, members of Class C help reduce the university's costs below the level they would be were full-fledged faculty to do the work done by the TAs. These tend to be part-time jobs.

Class D. The science or engineering graduate student who serves as a research assistant in return for financial assistance is playing the most important student role in the university labor market. The student not only helps reduce the university's costs, by working long hours at low pay (i.e., the stipend), he or she plays a key role in sustaining the flow of research, which maintains the flow of grants, which keep the graduate departments operating.

In fact, to many of us outside academe, the graduate research assistant or postdoctoral student looks very much like a highly-skilled, low-paid worker. This is a line of reasoning rejected by the university, which has good reason to want the rest of the world to view financial assistance to the undergraduate, as well as to the postdoc, as two parts of a seamless web.

Sweatshop U.?

What with so much sensitivity abroad in administration buildings these days, it's almost refreshing to hear of one university willing to come out and act like an old-fashioned evil corporation. Northwestern, upon learning...that one of its professors had been accused of forcing a visiting Chinese scholar into indentured servitude, responded with lofty indifference. "Interaction between visiting scholars and members of the faculty is largely a decentralized function," sniffed its Office of University Relations...

The tale of the indentured scholar began... when Lang Y. Xia, a male lab supervisor at Northwestern, wrote to Zeng Li Yang, a female neurophysiological technician in Beijing, and invited her to spend three years in his lab researching synthetic opiates. Xia, 54, had been Yang's teacher... until he moved to Chicago to work with Northwestern psychology professor J. Peter Rosenfeld. Yang, 40, accepted the offer and Rosenfeld confirmed the deal in writing.

Yang--who speaks no English--says she was promised $12,000 a year... and that is the salary stipulated in both the statement filed by Northwestern for her visa and the letter Rosenfeld sent her in China. Xia says that Yang agreed to work for nothing in exchange for her passage, and nothing is what she was paid. She ended up chopping vegetables in a campus cafeteria for less than $5 an hour while working fifty to sixty hours a week...at the lab...exploitation of Asian students has reportedly become common on American campuses, according to Lee Maglaya, head of the civil rights, race relations and advocacy program for Chicago's Asian Human Services agency...

from an article by Eric Zorn in
Lingua Franca, September/October 1993

Many talented and otherwise qualified people, particularly the native born, do not want to play the role of Class D student, but it is acceptable to many foreign-born graduate students, hence there are large numbers of them providing a hidden subsidy to the university.

Funding the Graduate Students. The process of providing financial assistance to graduate students is different from that used with undergraduates. Graduate students are regarded as adults, and there is less of an assumption that parental help is to be expected. Further, unlike most undergraduate situations where all bright and needy students compete on a university-wide basis for scholarships, funding at the graduate level is often tied to the department where the graduate student wishes to enroll.

At the graduate level in the leading research universities, full funding is often nearly automatic for everyone accepted by a science or engineering faculty. In many of these institutions, the university insists that departments assure full funding for every student the departments propose to admit. There are less affluent institutions that accept graduate students in science and engineering without full funding; they are, of course, placed in a difficult competitive position for attracting good graduate students.

"Full funding" is in the eyes of the beholder; a stipend of $12,000 a year plus tuition may meet the university's definition, and be attractive to overseas graduate students, but not to Americans. Or the stipend that once looked so good from Calcutta, proves to be less enticing after a couple of years in Boston. Or the same dollar level will produce a more comfortable student life in Blacksburg, Virginia, for example, than in Boston.

Not all foreign-born graduate students, of course, receive full funding, by which we mean a plausible stipend plus tuition and all other university costs. The range is a wide one, from the graduate student who personally pays for everything, (tuition, room and board, and fees) to the fully-funded person at an affluent research university. We encountered a number of steps along the continuum, including these: at an Appalachian institution (which specializes in MS degrees in electrical engineering for people from the Near East), tuition is paid by the state, and the stipend comes to $100 a month; at a somewhat larger Southern campus, tuition is waived for foreign-born TAs and RAs, and monthly stipends range from $600 to $800.

No matter how generous the university, there are four funding variables: the role of entrance fees; the inherent greater expense of supporting the foreign born than the native born; the different techniques universities use to fund graduate educations; and loans.

Entrance Fees. Only at the very beginning of the relationship is there a flow of money *to* the United States as a result of the interest of a would-be foreign-born graduate student in one or more American universities. The prospective student has to pay the previously mentioned ETS for taking the TOEFL and the GRE ($35-$42 is the basic overseas fee for the former and $65-$70 for the latter[88]). In addition prospective students must pay filing fees with each university where they apply; these fees range from of $50 to $100.

Since more than 100,000 GREs were given overseas in 1991-1992 and more than 550,000 TOEFLs (most of which are *not* for would-be graduate students), we estimate, using the mid-point in the registration fees, that ETS has overseas sales for these two products of $28,117,000. This is one area where foreign students are clearly helping America's balance of payments.

While these ETS and filing fees are significant to many native-born applicants, and less so to others, they are major obstacles to many applicants, particularly in PRC. With this in mind, the Department of Chemical Engineering at the University of Delaware goes through a two-stage applicant-screening process. The first stage screens out the weakest applicants, and Delaware's $50 fee is applied only in the second and decisive stage of the process. "We wrestled long and hard with this issue," Andrew Zydney, formerly chair of the Department's Graduate Admissions Committee, told me, "we were very worried about how much $50 means in much of the Third World, but if we did not retain the fee we would be overwhelmed."

The Graduate School at Dalhousie University, at the moment, waives its application fee of $30 (Canadian) when appropriate hardship evidence is submitted, but is aware that it is one of the few North American educational institutions to take this stand.[89]

Larger Grants for the Foreign-Born. Because of the high cost of air travel and the extremely limited incomes of some overseas applicants and their families, full funding of a foreign-born graduate student, at a given institution, is generally more expensive than full funding for

---

[88] The $35 TOEFL is for a Saturday test (when costs are lower) and the $42 is for Fridays; students may pay for the TOEFL in local currencies in many countries. GRE fees are $65-$70 in Japan, India, Korea, Taiwan, and Canada.

[89] We were informed during interviews at Dalhousie's Graduate School that one PRC graduate student had used a part of his stipend to establish a small fund in his PRC hometown to help other would-be overseas students meet these costs.

an otherwise comparable native-born student. Sometimes we were told that the sponsoring university had to pay for passport and visa fees as well as travel.

Different Funding Techniques. The previously mentioned Ronald Ehrenberg of Cornell University has identified four financial support patterns at the graduate level: fellowships (requiring no work for the institution), research assistantships, teaching assistantships, and other sources of support. He then explored the records of Cornell and ran a statistical analysis to see if different patterns of support correlated with different outcomes in terms of completion percentages or time-to-completion. He found that holding a fellowship or being a research assistant produced slightly better outcomes (particularly in terms of completion rates) than the other two patterns.[90] He theorized that perhaps those winning fellowships had academic skills superior to their peers and that working as an RA, rather than as a TA, brought the graduate student in closer contact with the professor, as the latter had a vested interest in the RA doing the research well. It happens that while new doctorate holders generally were about as likely to be TAs as RAs (46.3% and 45.9% respectively), those on temporary visas were more likely to work as RAs (64.8%) than as TAs (52.0%), while the reverse was true for citizens.[91] Perhaps the foreign-born are better researchers; certainly their command of English is less good, on average, than that of citizens, making them less useful as TAs. But whatever the universities' motivations, the foreign-born graduate students received their funding in a slightly more helpful manner than the native born.

How funds reach the graduate student can be important in other ways as well. Sometimes, we were told, a university will decide that Department Y simply must take on some additional students with certain characteristics; to facilitate this the university promises, out of

---

[90] Ronald G. Ehrenberg and Panagiotis G. Mavros, "Do Doctoral Students' Financial Support Patterns Affect Their Times-to-Degree and Completion Probabilities?" Working Paper No. 3, Institute for Labor Market Policies, School of Industrial and Labor Relations, Cornell University, May 1992. The studied graduate students were in the Departments of Economics, English, Mathematics, and Physics.

[91] *Summary Report 1991*, Table A-4; the same table in the 1990 report shows the same relationships.

university, not departmental, funds to support the applicant in question. The department agrees. The student arrives but may not succeed, as there is no single professor with a strong interest in the dissertation; thus the student has an orphan fellowship, outside the networks operating within the department, and suffers thereby.

The Matter of Loans. One of the reasons why foreign-born graduate students are less likely to use their own resources and less likely to be in debt than native-born students relates to U.S. government and university policies on lending money to foreign-born students.

The U.S. Congress decided some years ago that the principal federal source of loan funds to graduate students, the Stafford Loans (named after the former Senator from Vermont), would be made only to U.S. citizens and permanent resident aliens. In one sense, this Congressional decision worked, in that, in 1991, the new PhDs reported these usages of the Stafford program: citizens, 30.9%; green cards, 16.3%, and temporary visas, 0.2%.[92]

Meanwhile, some universities have also decided not to make loans to foreign students, usually on grounds that incomes in their nations of origin are low and the likelihood of repayment is slight.[93] (No one seems to have noticed that a large percentage of at least the science and engineering graduate students linger in the U.S. in well-paid jobs.) The upshot of the no-loans for foreigners policy, of course, is not that the foreign-born students do not come; they do come and get grants instead of loans, as has been shown by the NRC data.

The Role of the Postdocs

A generation ago there were not many "postdoctoral appointees," to use the National Science Foundation's term; if you had a brand-new PhD in science or engineering, you sought a job in industry or a junior faculty position, and, if you succeeded, you moved directly from graduate school to the new job. That is no longer the case.

---

[92] *Ibid.*

[93] Interviews with Ms. Marianne Waterbury, Assistant Dean of Students, Princeton University, and with various foreign student advisers at the May, 1992 NAFSA convention in Chicago.

Many with new doctorates now spend one or more years as postdocs; and, as we show subsequently, the foreign born are much more likely to do this than the native born.

Postdoc positions are largely in academe, although there are some in quasi-academic government agencies like the National Laboratories. The positions are usually short-lived, often for two years, and often nonrenewable. The typical work assignment is research, usually in the subfield already selected by the person taking the postdoc position. Again, as with the dissertation research, there is supposed to be a match between the postdoc's specialty and the interests of the employing agency. Often a postdoc will seek an appointment with a professor who is a leader in the subfield of interest; with luck he or she will find not only intellectual stimulation but a mentor who can help in the future.

In the eyes of the universities, the postdoc is lucky to have the assignment; it gives him or her a chance to work with a professor of his or her choice, and to continue learning. The universities view the postdocs' stipends as benefits distributed to the deserving and not as wages. To most young PhDs, a postdoc's position is a holding operation; an opportunity to pursue one's own line of research while looking for a *real* job; a postdoc's slot is not a real job.

An outsider might see the postdoc (at least in science or engineering) as a highly-educated junior professional whose work allows the research enterprise or institution to continue to attract grants; this, in turn, permits it to survive and perhaps prosper. A postdoc is often working as a relatively high-level player on a research team that is headed by a professor (the grant writer) and includes one or two postdocs, several graduate students, and perhaps others (such as some part-time undergraduate workers, or one or more members of the university's research staff).

The wages for postdocs range widely, even wildly; many seemed to be getting, in 1993, as little as $18,000 a year; in some cases the stipend can be close to $40,000. Although the government funds many postdocs and collects a lot of data on doctoral students and postdoctoral appointees, we were able to find only the most fragmented information on postdoc wages.

What we thought was the most likely source of the information, the Department of Labor, was not. Although institutions seeking to hire postdocs as H-1B nonimmigrants are required to file the previously mentioned Labor Condition Application (LCA) with the

Employment and Training Administration, the wage data from these forms are not captured in a useful way. DOL uses a three-digit, *Dictionary of Occupational Titles* coding system, which lumps the postdocs together with full professors and deans. That neither the Labor Department nor the National Science Foundation keeps useful data on postdoc wages, though both have serious program responsibilities with this population, suggests both a lack of agency interest in the issue and the postdocs' total lack of anything approaching a self-help or advocacy organization.

We did find three partial data sources on postdoc wages: a National Research Council survey that covered 853 postdocs in 1991;[94] a University of Nebraska survey that covered 2,388 of them for the academic year 1991-1992,[95] and NIH data on 6,429 postdocs working on NIH grants in FY '90.[96]

Collectively the three data systems provide information on 9,640 postdocs in the early 1990s, when there were 25,000 or so postdocs in U.S. science and engineering.[97] Unfortunately the three surveys used different techniques and produced different results. The NRC study used self-reported wages while the other two depended on employer information. Further, the major research universities (the Ivies, Stanford, MIT, etc.) did *not* respond to the Nebraska survey.

The NIH population is largely in the life sciences, but these findings are probably a good reflection of the low wages paid to postdocs in those fields. The NIH survey covered postdocs paid by NIH grantees, not by NIH directly. NIH found that the median

---

[94] Unpublished NRC data provided by Alan Fechter of NRC.

[95] Office of Research and Graduate Studies, University of Nebraska-Lincoln, *National Survey of Graduate-Assistant Stipends, Graduate Fellowships, and Postdoctoral Fellowships, 1991-92*, University of Nebraska-Lincoln, Lincoln, Neb., 1993, Section IV (no page numbers).

[96] Office of Science Policy and Legislation, NIH, *Staffing Patterns of the National Institutes of Health Research Grants*, NIH, Bethesda, Md., 1993, p. 19.

[97] An NSF publication, which does not include any salary information, reported 23,018 postdocs employed in doctorate-granting institutions in 1991. See J.G. Huckenpöhler, *Selected Data on Graduate Students and Postdoctorates in Science and Engineering: Fall 1991*, (NSF-335) NSF, Washington, D.C., 1992, p. 45.

annual income for someone working year-around as a postdoc came to $13,935 in 1983, increasing to $19,171 in 1990.[98] When inflation is taken into account, there were only minuscule wage increases (one tenth of one percent a year we estimate) between 1983 and 1990.

To provide a frame of reference, Exhibit Twelve shows how weekly earnings of the postdocs in these three groups relate to those of other (male) workers in 1990. Although NIH-funded postdocs are likely to be in their late twenties or early thirties, with something like 22 years of education and training, they make less than workers on an assembly line. In fact, there are only three broad categories of workers — laborers, other service workers and farm workers — who earn less than the $369 a week that NIH postdocs earn. The other, and smaller, surveyed groups of postdocs have higher weekly wages than the NIH-funded ones.

Another measure of the adequacy of the NIH postdoc wage levels can be secured by comparing them to those that the Internal Revenue Service officially recognizes as those of the working poor. The U.S. Tax Code provides Earned Income Credits (EIC), in a sense a negative income tax, to families with children where the family's income is below a stipulated level (which rises yearly). In 1990, when the median NIH stipend was $19,171 a year, the maximum earnings to receive EIC benefits was $20,264. There is no indication that NIH, or any other government agency, ran a public information program to let the postdocs know that they were badly enough paid to qualify for that benefit.

Postdocs in physics seem to earn more than those in life sciences, according to Dr. Aylesworth; he said that his experience as a postdoc in physics (when he was paid considerably more than the NIH figures quoted above) and that of his peers in that discipline suggested that a management-perceived sense of a tight labor market in physics had increased postdoc wages there.[99]

---

[98] Office of Science Policy and Legislation, NIH, *Staffing Patterns of the National Institutes of Health Research Grants*, NIH, Washington, D.C., 1993, p 19. Since the length of the postdocs' assignments vary, sometimes covering just the academic year, NIH calculated what the stipend would have been had a full twelve months been worked; if anything, this estimation technique may create an annual payment rate figure that is larger than what a postdoc actually earns in a 12-month period.

[99] Telephone conversation with Dr. Aylesworth, Sept. 13, 1993.

Exhibit Twelve: How Postdocs' Weekly Wages Compare with Those of Male Workers Generally, 1990: BLS and Survey Data

(median weekly wages in $)

Occupation	Dollars per week
Exec., Admin., Managerial	$742
Professional Specialty	$720
Tech. and Related Support	$570
Post-Docs, NRC 1991 Survey (n = 853)	**$535**
Male Workers (25+)	$514
Other Precision Production	$510
Sales	$505
Male Workers (all ages)	$485
Construction Trades	$480
Protective Services	$477
Mechanics and Repairers	$477
Admin. Support, incl. Clerical	$440
Transportation and Material Handling	$418
Post-Docs, U. of Neb. Survey (n=2,388)	**$400**
Machine Operators, Assemblers and Inspectors	$391
Post-Docs, NIH Grants (n = 6,429)	**$369**
Handlers, Equipment Operators, Helpers and Laborers	$308
Other Services	$273
Farming, Forestry, Fishing	$263

Note: Weekly earnings data are for male workers except for postdocs, which are for both men and women. Most postdocs in science and engineering are men.

Sources: Weekly earnings for male workers generally, *Statistical Abstract of the U.S. 1992*, table 654; weekly earnings for postdocs calculated from three sources cited in the text.

> **The Perpetual Postdoc**
>
> She was easy to notice at the American Chemical Society convention's job bank (in 1992). A citizen of Bangladesh, she wore the traditional floor-length, long-sleeved outfit and the gauzy veil of the devout. She held a PhD in biochemistry from an American university. She supported her nonimmigrant husband, a graduate student, and her two children. She was nearing the end of her third consecutive postdoctoral appointment, at her third U.S. university, and was looking for a real job.

One reason for the continuing low wages for postdocs is the rapid expansion in this work force in recent years, with the foreign born accounting for about 72% of the growth. Exhibit Thirteen shows that the number of postdocs increased by two-thirds between 1980 and 1991 and that this is rapidly becoming a population dominated by the foreign born. While 38% of the nation's postdocs were non-USCs in 1980, more than 52% were non-USCs by 1991. In the science fields (as we define them), there were almost as many non-USCs as USCs, and in engineering, where smaller numbers are employed, the postdoc population was more than two-thirds non-USC. Alien postdocs are more numerous in some fields than others, and the variations reflect those in the distribution of PhDs earned in the U.S. For example, in 1991, these data were reported:

Discipline	% Non-USCs[100]
Science, total	**49.9%**
Mathematics	60.9%
Physics	60.2%
Biology	46.0%
Computer Science	43.7%
Earth Sciences	35.8%
Engineering, total	**68.8%**
Mining Engineering	94.7%
Chemical Engineering	71.1%
Electrical Engineering	67.6%

---

[100] See Exhibit Thirteen for the source of these data; there were, incidentally, only 19 mining engineering postdocs in 1991, one USC and 18 non-USCs.

Exhibit Thirteen: The Postdoc Labor Force, 1980-1991; Growing Rapidly With an Even More Rapid Increase Among the Foreign Born

Category	1980	1985	1991	Increase 1980-91
Total Science and engineering	12,482	15,092	20,931	68%
Non-USC	4,772	6,264	10,831	127%
USC	7,715	8,828	10,100	31%
% Non-USC	38%	42%	52%	
Science	11,415	13,898	18,978	66%
Non-USC	4,119	5,505	9,487	130%
USC	7,296	8,393	9,491	30%
% Non-USC	36%	40%	49.9%	
Engineering	1,067	1,194	1,953	83%
Non-USC	653	759	1,344	106%
USC	414	435	609	47%
% Non-USC	61%	63%	69%	

Source: J.G. Huckenpöhler, *Foreign Participation in U.S. Academic Science and Engineering*, National Science Foundation, Washington, D.C., 1993, table A-24.

Note: the totals and percentages shown above for Total Science and Engineering and for Science are different from those in the source table because we backed out about 1500 postdocs each year whose fields were agriculture, social science, and psychology.

Presumably bearing in mind both their need to move from institution to institution and their wages, Daniel S. Greenberg, while discussing forecasts of further declines in research and development funding, wrote of "the migratory workers of modern research, the hapless platoons of postdocs."[101]

Joining the Faculty

It is well known that foreign-born S/Es are playing a larger and larger role in science and engineering faculties in America's universities. Clearly it is an able population and one playing a growing role in the receipt of PhDs and in securing postdoc slots, thus swelling the traditional pipeline toward faculty appointments.

Throughout much of this research we were also convinced that one of the reasons for the growing faculty role was not only because the foreign-born PhDs were growing in numbers, but because a *disproportionately large* percentage of them went to work for universities. As one observer put it, "Foreign citizens, as well as naturalized Americans, are somewhat more likely to be employed in colleges and universities than in industry."[102]

Some of the most widely printed numbers on this point can be seen in Exhibit Fourteen; they deal with a fairly small population, assistant professors of engineering aged 35 and younger. A glance at the exhibit indicates that — like so many statistical reports one sees in this field — the number of foreign born is increasing rapidly.

A more careful look reveals the most volatile flow of numbers seen in any of the exhibits in this report and suggests caution. By definition, the people covered by this segment of NRC's Doctoral Recipients' Study (a sample taken every two years) are not a stable population; new assistant professors under age 35 are appointed all the time; existing assistant professors leave academe (for industry), are promoted within academe, or move from the non-USC to the USC column through naturalization. Further, and perhaps more

---

[101] Daniel S. Greenberg, "A New, Deeper Pessimism Sweeps Over Science," *Science and Government Report*, Washington, D.C., August 1, 1993, p. 3.

[102] Betty M. Vetter, "Foreign Citizens Among U.S. Scientists and Engineers," Commission on Professionals in Science and Technology, Washington, D.C., 1992, p. 11.

Exhibit Fourteen: Civil Status of Assistant Professors of Engineering, 35 Years of Age or Younger, 1973-1989

Year	USCs	Non-USCs	Unweighted Number
1973	87.6%	12.4%	209
1975	89.9%	10.1%	179
1977	87.4%	12.6%	157
1979	77.1%	22.9%	70
1981	72.1%	27.9%	76
1983	46.3%	53.7%	74
1985	50.5%	49.5%	112
1987	49.6%	50.4%	252
1989	58.9%	41.1%	375

Source: 1973-1985 percentages from Office of Scientific and Engineering Personnel, National Research Council, *Foreign and Foreign-Born Engineers in the United States: Infusing Talent, Raising Issues*, NRC, Washington, 1988, table A-13; other data over the telephone from NRC's Survey of Doctorate Recipients.

Note: see the text for a comment on the volatility of these numbers.

---

important, the volume of respondents has varied on the order one to five over the years.

The Doctoral Recipients Study, however, has other, more broadly based data dealing with this subject. These data should be examined as well; they show that as the foreign-born PhDs grow older and acquire citizenship, they become less likely to be academic employees than the native-born.

Percent of Science and Engineering PhDs Employed by Educational Institutions, 1989[103]

Civil Status	%	No.
Native-born USC	51.6%	320,372
Natz. USC	45.7%	47,905
PRA	58.7%	22,576
Nonimmigrant	65.7%	5,299

There appears to be an explanation that fits both the steadily rising percentage of non-USCs as under-35 assistant professors of engineering and the more broadly based data shown above. That pattern may be that both a postdoc position and subsequently a junior faculty appointment are stepping stones for the newly minted foreign-born PhDs as they find their way around the American economy or decide to return to their homeland. As we show later, there is a similar pattern in earnings, with a direct relationship between a closer bond to the U.S. (i.e., naturalization) and higher earnings. As time passes and as nonimmigrant S/Es become first permanent resident aliens and then naturalized citizens, there is a movement away from academe and toward an increase in wages.

What we do not know and cannot learn from currently available data sources, is the extent to which this movement away from academe over time is a successful integration of foreign-born S/Es into U.S. industry, or alternatively, a successful integration by the many as well as the emigration of a few of the less successful ones. Another possibility may be that many of the naturalized citizens are from a prior generation, largely European, who were less drawn to academic life than the current generation of foreign-born S/Es. To some extent, all of these factors may be at work.

We turn from this statistical analysis by an outsider to an intriguing insider view of the role played by the foreign born on science and engineering faculties. We encountered this strong suggestion, in a National Research Council report, that there may be some institutional tendencies to favor the foreign born in the faculty-hiring process:

---

[103] The data are from Joseph Gannon and Marjorie Lueck, *Characteristics of Doctoral Scientists and Engineers in the U.S.: 1989*, National Science Foundation (NSF 91-317), Washington, D.C., 1991, tables 16 and 25.

> Salaries paid to assistant professors of engineering have increased dramatically in recent years and are now comparable with, or superior to, those paid by industry, when allowance is made for summer salary supplements and consulting income. In view of this dramatic improvement in salaries at major universities, it is not surprising to find large numbers of applicants for faculty openings at the research universities. Quoted numbers are 50 to 200 or more for each widely advertised position. The question arises why the normal academic selection procedures, when applied to openings for which there are so many potential applicants, have yielded a foreign and foreign-born component in excess of 50 percent, a component that is probably increasing.
>
> The answer may be found, at least in part, in faculty preferences for people with high analytical ability and/or particular skills in utilizing advanced instrumentation techniques and relative de-emphasis of what may be called the art of practical engineering as compared to engineering science. Thus while maintaining "quality" in academe according to current preferences, the "character" of engineering education may well be changing dramatically. We believe that a careful assessment of the likely long-term impact of these changes forms an appropriate and urgent subject for evaluation.[104]

The seven signatories of this document, including Jules LaPidus, President of the Council of Graduate Schools, and Dorothy S. Zinberg, of Harvard's Kennedy School, are hardly alarmists about foreign-born penetration of U.S. systems, as the balance of the report indicates, but the carefully worded passage above suggests an extremely strong concern about the increasing foreign-born role in engineering faculties. The same set of thoughts, particularly the implication that U.S. citizens may be discriminated against in hiring, might be discussed in more inflammatory terms if written by others.

The Foreign Born in Industry: The Older, Larger Pattern

There are, as mentioned earlier, two different industrial patterns regarding the employment of foreign-born S/Es, which we have characterized as the immigrant and the nonimmigrant patterns. The first is the one described extensively earlier, in which foreign-born S/Es come to the U.S. as graduate students and then opt to stay; in the course of this pattern, over the years they typically move from nonimmigrant to immigrant to naturalized citizen status. Others in

---

[104] *Foreign and Foreign-Born Engineers in the U.S., op. cit.*, pp 16-17.

this pattern come as immigrants after finishing their professional education overseas. All move through the pattern one at a time, making their own decisions, working their own networks. Most of the statistics cited earlier show that this is a growing population, though one must often rely for confirmation of the trend on such indirect measurements as degrees granted and citizenship achieved.

The second pattern, that of nonimmigrant movements totally unconnected with higher education, has increased rapidly in the last couple of years and has secured its own 15-minutes of fame, a highly critical segment on the CBS-TV program "60 Minutes" on October 2, 1993. The opponents of this pattern in the media and the engineering associations regard it as a technique for laying off American workers and replacing them with less expensive foreign ones. It, unlike the first pattern, involves the movement of groups of workers organized by middlemen, usually from the country of origin.

These two patterns are so different from each other, and with such different consequences, that we will describe them separately.

The Immigrant Pattern: Numbers and Characteristics

Chapter One described in some detail the growth of the foreign-born scientist and engineer population in recent years, the greater incidence of PhDs among them than among native-born scientists and engineers, and the foreign-born preference for the physical sciences and engineering, as opposed to the life sciences.

The immigrant pattern is such that it tends to be documented much better than the nonimmigrant pattern. If one is a long-time resident of the U.S. as opposed to a guest worker, if one is a PRA as opposed to a nonimmigrant, if one secures a degree here, rather than abroad, one gets counted in the systems described earlier.

The Immigrant Pattern: Sectors of Employment

In the immigrant pattern, foreign-born S/Es are distributed among sectors of the economy similarly to native-born ones. In fact, the variable of interest regarding sector of employment is the degree, not birthplace. At the PhD level, about half of both groups work for educational institutions, about a third work in industry, and the rest work for government and in the nonprofit sector.[105] When all S/Es are included, not just PhDs, the distribution by sector is somewhat different. The leading sectors in the 1982 post-census study showed the following distributions of native-born and foreign-born S/Es:

Sector	Total	% Native Born	% Foreign Born
Industry	688,826	82.8%	17.2%
College/Univ.	116,931	82.7%	17.3%
U.S. Govt. (civ.)	105,326	89.2%	10.8%
State, Local Govt.	59,241	83.6%	16.4%
Total[106]	1,111,317	83.1%	16.9%

It is worthy of note that there are significant differences in the incidence of employment of the foreign born at different levels of government; ironically the federal government (which sets immigration policy) is a much less welcoming employer than state and local governments (which do not).

Two sets of variables appear to have been at work here when the 1982 data were collected: citizenship requirements and wage levels. Lingering federal requirements for citizenship, even in the nonmilitary agencies, apparently discouraged the hiring of non-naturalized foreign-born S/Es, while higher salary levels attracted more native-born S/Es into federal employment. Conversely, state agencies, with lower salaries to offer — but few, if any, citizenship requirements and badly needing scientific talent — were more likely to hire foreign-born S/Es than federal agencies.

The Immigrant Pattern: Occupation

Within the traditional employment pattern, there are interesting differences in the kinds of work that engage the four classes of interest, with the native born being more involved in management

---

[105] Gannon and Lueck, *op. cit.*, table 25.

[106] Includes other, smaller categories as well as those shown. From Finn, *Foreign National Scientists and Engineers, op. cit.*, table A-20.

> **Foreign Born S/Es as State Employees**
>
> We visited a unit of state government, part of Maryland's Health Department, that deliberately reaches out to foreign-born S/Es as a part of its personnel policy. This laboratory's function is to test various fluids: drinking water, milk, sewage, etc.; most of its employees are chemists, and they are largely Black, or female, or both, and most are foreign-born.
>
> The lab director, a Ph.D. and a USC, said that the salaries offered were such (e.g., $24,000-a-year for a BS chemist) that he had very few native-born applicants and that the Americans applying often had marginal resumes. But, as is often the case in niches in the labor market, these professional vacancies could be meshed with the availability of newly-arrived foreign-born S/Es.
>
> The lab director had a very coherent personnel policy; he stayed in touch with refugee- and immigrant-serving organizations and let them know when the Maryland civil service structure was advertising jobs in his lab. He had thorough knowledge of the civil service rules and used that knowledge to keep his place staffed and his workers happy; and he used the state-supported training system to help his employees secure additional skills.
>
> He said that his new staff members often had only foreign credentials making it hard for them to land other jobs. "I get one to three years of good lab work out of them before they move on to better jobs -- and I provide them with U.S. work experience and U.S. credentials."   He said that federal agencies, like EPA, often raided his staff, offering them $10,000 a year more than he could pay.
>
> Did he offer his workers a green card as part of the compensation scheme? "No, but I would if I had to; most of my people already have the green card because of refugee status or because of a marriage."
>
> The statistics suggest substantial foreign-born science and engineering employment in state and local government; we have no idea if the process is always as deliberate, or the boss as thoughtful, as in this lab in Baltimore.

and the newest of the foreign born being heavily employed in research and development. In other words, the native born are more likely to be the bosses, and the non-USCs are more likely to work in the laboratory, as these data show:[107]

---

[107] See *Characteristics of Doctoral Scientists and Engineers, op. cit.*, table 25.

	Native-Born USCs	Natz. USCs	PRAs	Nonim.
Primary Work Activity				
Management	17.2	15.6	8.7	0.2
R & D	35.3	41.9	50.1	65.5

The data are for all doctoral S/Es, working in all employment sectors. Many of the nonimmigrants were presumably postdocs.

The disproportionately low foreign-born participation in science and engineering management has been criticized by a former member of the U.S. Equal Employment Opportunity Commission, Joy Cherian; Mr. Cherian is a native of India and a PhD. While he did not cite statistics in the material we have seen, Cherian used examples from the science and engineering environment in his charges about a "Glass Ceiling" being imposed on Asians:

> "If you go to the National Institutes of Health," Cherian said, "you will see a lot of foreign-born scientists, and many do not go beyond GS-14," about a middle-management level. He said the same applies to many private corporations. "You won't see many at the vice president or senior vice president level," he said.[108]

One corporate response to this matter is to create two career science and engineering pay ladders with equal levels, one for scientists and one for managers. The loftiest levels, above the top of the dual pay system, can be filled from either the management or the technical hierarchy, but presumably usually from the former.

The Immigrant Pattern: Wages

Given Commissioner Cherian's comments and Parlin's findings regarding corporate discrimination against the foreign born in science

---

[108] Frank Swoboda, "Foreign-Born, Too, Face 'Glass Ceiling' in Job Promotions," *Washington Post*, March 10, 1991, section H, p. 10. See also Commissioner Cherian's speech "National Origin Discrimination in the Workplace: a Glass Ceiling Issue For Asian Americans," delivered on June 1, 1991, at Roosevelt University, Chicago, during a conference sponsored by Senator Paul Simon's Asia American Advisory Committee.

> **Have Them Report in Writing**
>
> It was breakfast at an electrical engineering convention; the three of us were strangers until then. She, a young native of Florida, was a designer and monitor of construction for utility substations; he was a retired chief of engineering for a Texas utility. After she left, he began talking about his firm's use of foreign-born engineers.
>
> He said that his experience was pretty limited, as "not many of them find their way to Amarillo." He went on to say that "supervising someone with limited English takes more of my time, so I gave [the one foreign-born engineer] a job where he can use his talents and can report to me, largely, in writing." So he put the Chinese engineer to work revising the computer system.
>
> "She," referring to the Floridian, "has a job where she is talking to people all day, to contractors, to regulators, and to others in the utility; it would not be a good job for my guy."
>
> Her very presence at the convention--there were not many women--suggested that she was already on the management track; the Chinese computer whiz, however, had an assignment where he would have little chance to practice his spoken English. Their bosses' early expectations may make an impact on the kind of work they do for the rest of their lives.

and engineering hiring practices, we now turn to wage scales. We know that women, Blacks, and Hispanics in the work force are generally paid less for similar work than white males. Do foreign-born S/Es, similarly, get lower salaries than their native-born peers? The answer, at least in the immigrant pattern of foreign-born S/E utilization, apparently not only is "no," but also shows strong evidence that naturalized USCs in science and engineering earn a little *more* than their native-born peers, when other factors are held constant.

We are making a comparison here between the two largest of the four civil status groupings of interest, native-born and naturalized citizens. Not only are there many more naturalized USCs than non-USCs among the foreign born, as we have noted several times, but there are age and time-in-country factors that tend to produce lower earnings for the non-USCs than for the naturalized USCs (who have been in the U.S. longer). Age and years of corporate seniority are important variables in setting salaries anywhere, and particularly in the bigger, older, and more affluent companies.

There are many salary surveys within science and engineering; members of these professions are used to answering questions (usually anonymously) on this matter. Not all of the surveys pick up the variable of civil status, and some of those that ask the question do not use the variable in their tabulations. But we did locate three major surveys that provide data by immigration status.

The most recent of these data are the regular biannual salary surveys of the American Chemical Society (ACS), the results of which are shown in Exhibit Fifteen. Separating results for industry and academe, and for 1990 and 1992, we find in every comparison that:

- The naturalized USCs *always* earn more money than native-born USCs, (sometimes by a slight margin) and

- Native-born USCs *always* report higher earnings (usually by large margins) than non-USCs.[109]

The ACS study, however, dealt with only a single discipline, and only with those holding doctorates. Would broader studies show the same patterns?

One such survey was NSF's 1989 Survey of Doctorate Recipients, covering the sciences and engineering. It reported median annual (self-reported) salaries by the four civil status groups and crosstabulated them by dozens of variables, some of which are shown in the following table:

Variable	Civil Status			
	Native-Born USC	Natz. USC	PRA	Nonimgrt
All	$53,900	$60,100	$52,600	$40,700
Men	$55,800	$60,600	$53,900	$40,700
Women	$44,500	$48,900	$44,400	$40,600
White	$54,100	$61,700	$54,900	$42,100
Asian	$50,500	$58,500	$51,500	$42,300
Black	$48,700	$48,400	$51,000	n.a.

---

[109] The author is grateful to Dr. Joan Burrelli, Senior Research Analyst of ACS, for the special tabulations she ran for this study.

Exhibit Fifteen: Naturalized Chemists Earn a Bit More than US-born Chemists, 1990-1992: Results of Salary Surveys Among ACS Members Holding PhDs
($/year)

	Industry, 1990	Academe, 1990	Industry, 1992	Academe, 1992
Naturalized citizen	$68,752	$53,129	$72,945	$52,442
n = ( )	(857)	(269)	(362)	(113)
US-born citizen	$68,273	$45,199	$72,019	$48,940
n = ( )	(7778)	(3543)	(2459)	(1074)
Permanent resident	$59,674	$44,5654	$65,971	$42,736
n = ( )	(575)	(170)	(245)	(63)
Other Visa	$52,720	$32,777	$49,510	$38,598
n = ( )	(114)	(30)	(66)	(13)

Notes: Data shown are means of self-reported salaries; period covered is 12 months for industry, 9/10 months (the academic year) for those in academe. "Other Visa" is the ACS term for those with nonimmigrant visas. Sample sizes were different in the two survey years.

Source: unpublished data from the regular Salary Surveys of the American Chemical Society, Office of Professional Services, Washington.

So while men continue to earn more than women, and whites are better paid than either Asians or Blacks, native-born USCs earn less than the naturalized, except for a statistically insignificant $300 break for native-born Blacks over naturalized Blacks.

There were a total of 64 cross tabulations in this study;[110] an examination shows that in 51 cases, the naturalized USCs earned

---

[110] Gannon and Lueck, *op. cit.*, tables 32 and 33.

more than the native-born USCs. Three of the exceptions were of interest: (1) among the youngest of the respondents, those under 30, native-born USCs did slightly better than naturalized USCs; (2) the native-born computer science people earned $6,600 more than their naturalized peers; and, (3) along the lines of Cherian's example of NIH, native-born citizens working for the federal government earned $54,100 compared to $52,800 for the naturalized.

Unpublished data on the subsequent (1991) iteration of the same study again showed the same pattern, with the naturalized citizens reporting median incomes of $65,600 and the native-born reporting incomes at $60,700. Again, naturalized men earned more than native-born men, and naturalized women PhDs earned more than native-born women.[111]

We turned to yet another survey, the 1982 post-census, and found this analysis by Finn: "Foreign nationals earn salaries comparable to their U.S. counterparts. As a group foreign nationals tend to earn slightly lower average salaries, but this is largely because they tend to be younger and have less work experience than their U.S.-born counterparts... when a simple adjustment is made for the difference in age, the ratio of foreign national salaries to U.S. native salaries is reversed."[112] The post-census population providing these data consisted of all full-time employed male S/Es, not just those with PhDs.

One more example: The 1990 Census, covering all workers in science and engineering, male and female, PhD and no PhD, again showed higher earnings for the foreign born than the native born in science and engineering. This is a cruder measurement than most of those previously described. On the one hand, it makes no allowance for time in the country as the comparison of only the naturalized PhDs versus the native-born PhDs does; on the other, it compares a population with a high proportion of PhDs (the foreign born) with a population having a lower proportion of PhDs (the native born).

Lee Bouvier's mining of the 1990 Census data shows the following

---

[111] National Science Foundation/SRS, 1991 Survey of Doctorate Recipients (National Research Council), table 32.

[112] Finn, *Foreign National Scientists and Engineers in the U.S. Labor Force, 1972-1982, op. cit.*, page 14.

median income data (reflecting 1989 incomes):[113]

Occupation	Foreign Born	Native Born
Engineers	$40,352	$39,640
Math./Computer Science	$38,105	$36,640
Natural Science	$33,421	$32,817

We did not explore the glass ceiling question, though we have the impression that if an Asian scientist wants to be a CEO he sets up his own firm (as did the late Dr. An Wang, of Wang Laboratories). But if the question is does America pay its foreign-born citizens less than its native-born citizens in science and engineering, the answer is clearly no.

Why the generally higher salaries for the naturalized among the PhDs? This question cries out to be researched in depth, but is beyond the scope of this study. In the meantime, here are four possible reasons: (1) perhaps it simply reflects the tougher screening process imposed by graduate schools on would-be graduate students among the foreign born than on the native born; (2) perhaps there is differential post-PhD outmigration among the foreign born, with the ablest (or most aggressive or luckiest) staying in the U.S. and the lesser ones going home, leaving an elite subpopulation among the foreign-born PhDs who stay in America to compete with the *entire* native-born PhD population; (3) perhaps the scientific ethos where these people work encourages a more rational, less ethnocentric approach to salary setting than found in some other parts of the American economy; and (4) perhaps it is much simpler: the foreign born simply work harder, in the grand tradition of the American immigrant's work ethic.

One of the mysteries of this field is why, despite the statistical data on the point, no writer, other than Lee Bouvier, has previously pointed out that naturalized citizens among S/Es consistently report higher average earnings than native-born citizens, just as no one

---

[113] Bouvier and Simcox, *op. cit.*, table 3.12.

seems to have noticed that the foreign born apparently receive more generous funding in graduate school than the native born.

The Foreign Born in Industry: The Newer, More Controversial Pattern

The Nonimmigrant Pattern: the Use of B-1s and H-1Bs

The nonimmigrant pattern of bringing foreign-born S/Es to the U.S. involves group movements of foreign-born workers, low wages, and, so far, a fairly narrowly focused use of these workers. This pattern has two variations — not always treated separately by the media — with one being more exploitive than the other.

The more exploitive of the two variations is probably illegal but clearly hard to control. It is the use of the B-1 visa program to bring groups of very low wage professional workers, usually software programmers, to the U.S. Its critics want the U.S. Government to make use of current laws to eliminate the program.

The other nonimmigrant program makes use of the H-1B visa to bring groups of relatively low-wage science and engineering workers, usually in software, to work for major U.S. corporations. Critics contend that changes are needed to provide greater protection for both the H-1Bs and for competing U.S. workers.

Supporters of both programs argue that they are needed to provide U.S. software firms with inexpensive talent that will allow American companies to cope with worldwide competition;[114] they sometimes make the point that, given modems and satellites, software can be written anywhere in the world, and that the B-1 and H-1B programs allow the work to continue to be done in the U.S., even though performed by nonimmigrants. Supporters of the H-1B program make the further point that the practices are within the

---

[114] For a strong defense of a welcoming policy, generally, vis-a-vis foreign-born S/Es, written from the point of view of the immigration bar, see Gary Endelman and Robert Loughran, "The Reality of Reliance: Immigration and Technology in the Age of Global Competition," *Immigration Briefings*, July 1993, Federal Publications, Washington, D.C., 1993. For a thoughtful critique of the B-1 program in science and engineering, see Demetrios G. Papademetriou, "White Collar Nonimmigrant Workers: B-1-in-Lieu-of-H-1B Issues," Carnegie Endowment for International Peace, Washington, D.C., 1993.

bounds of (current) U.S. laws and regulations, which, in a narrow sense, is probably correct.

It may be useful to describe the gatekeeping operation used in these two programs, compared with the rather more careful ones described earlier, before discussing the use of these programs in American industry.

B-1 visas are the second oldest of the nonimmigrant visas in the U.S. immigration business — only the A visas for foreign diplomats are more senior. B-1 visas were (and are) designed to make it possible, and relatively easy, for foreign commercial travelers to come to the U.S. to buy, sell, negotiate, attend conferences, and the like. They are issued overseas at U.S. consulates without any need for comment or consent by an American institution (other than the State Department). There were 2.7 million admissions of B-1s in FY '91.

H-1B visas are issued less frequently; there were 118,000 admissions of people with H-1, the predecessor visa, in FY'91, and 61,591 approvals of H-1B petitions in FY '93.[115] Admissions and approvals are different concepts (an approval can be for more than one person), and the current H-1B category covers a narrower set of people than the old H-1, but these totals provide an order-of-magnitude contrast to the B-1 activity. The H-1B visas are for people who plan to work in the United States; in earlier times they were regarded as temporary workers of "distinguished merit and ability." Now, although a fairly similar population is using the system, they are regarded as workers in "specialty occupations," primarily the professions. The three agencies with formal involvement in the H-1B process, are, as noted earlier, agencies that historically have approved the vast majority of the applications presented to them.

The B-1 Operation. The B-1 program works as follows in U.S. science and engineering: A consulting firm, probably with its feet planted firmly in both the U.S. and in the country where the workers are to be recruited, approaches a U.S. company.[116] The consulting

---

[115] "INS Approved 61,591 H-1B Petitions Last Fiscal Year," *Interpreter Releases*, Nov. 1, 1993, pp. 1438-1439.

[116] Most of the documentation on the use of the B-1 visa by consulting firms is journalistic; see, for example, John Eckhouse, "Programmers Losing Out: Immigration Loophole Lets Foreigners Take U.S. Jobs," *San Francisco Chronicle*, July 6, 1992; Al

firm offers to provide on-site programming services at a fully-loaded rate of say $30 per hour, perhaps $20 per hour less than the competition (which does not use the B-1 program). The consulting firm lands the contract.

It then signs up a group of programmers with the needed skills and sends them to the U.S. consulate to secure B-1 visas. The programmers tell the State Department officials that their employer is sending them on a business trip to the U.S., and routinely the visas are granted.

Once in the U.S., the programmers report to the U.S. software user (or developer), sit down at a keyboard in offices in the U.S., and do the work the consulting firm had agreed to do. The programmers are paid by the overseas consulting firm, sometimes as little, the news reports say, as $250 a month; the payments are made in the homeland currency and are deposited in the worker's homeland bank account. The consulting firm houses the programmers — the norm seems to be four of them in a two-bedroom apartment — and gives them a food allowance. It is often estimated that the consulting firm can pay the worker and cover the worker's rent and food for about $1650 a month, while receiving $1,200-$1,600 a *week* for the worker's services.[117] The programmers often work more than 40 hours a week, and, of course, the consulting firm does not pay any FICA or other U.S. taxes. As Papademetriou points out in his Carnegie paper, while the consulting firm is highly likely to know that is breaking U.S. law by these practices, the U.S. client may not realize it.[118]

It is fairly clear to all players — the State Department, INS, and the critics of the program — that an alien should not sit down and do a U.S. job while holding a B-1 visa. The central problem is how does a gatekeeping system as fragile as this one and as distant from the

---

Kamen "Visa Loophole Seen Costing U.S. Workers Computer Jobs," *Washington Post*, November 9, 1992, page A4; Charles Child, "Computing An Unfair Advantage? Critics Say Companies Misuse Visas," *Crain's Detroit Business*, July 29, 1992. The tone of the articles, incidentally, is accurately reflected in these headlines.

[117] The $1,650 per month estimate is from the previously cited *San Francisco Chronicle* article; the estimate does not include the other major cost to the body shop, the airfare to the U.S.

[118] Papademetriou, *op. cit.*, p. 11.

American labor market enforce the rules?

At this writing State and INS have been struggling for more than a year to write, clear, and then impose new regulations that would facilitate enforcement efforts against these abuses of the B-1 program.[119] The prospects for effective reform are not promising; consular officers have more pressing worries, like barring terrorists, and INS has never been particularly interested in fighting labor exploitation by American corporations. Perhaps the best hope for American programmers is that executives of major American firms will decide that they do not want to be chased — on network television — by "60 Minutes" reporters trying to grill them on the subject.[120]

The H-1B Operation. There are similarities and differences when the two nonimmigrant programs (B-1 and H-1B) are compared. Both provide on-site services to U.S. corporations, both recruit skilled workers in the homeland en masse for big projects, both use nonimmigrant programs, and both are said to bring in workers at lower wages than those paid to U.S. workers. Further, both are best known as operating in the software industry.

The difference is that it is perfectly legal to hire an H-1B for the U.S. labor market. The controversy is over the wages paid to the workers and the amount of labor market regulation that is needed to prevent displacing U.S. workers and depressing of U.S. wages. Another potential difference, not yet encountered, relates to the numerical limit, now set at 65,000 for annual admissions of H-1B visa holders; at the moment the limit has been approached, but not yet reached. There is no numerical limit on B-1 visas.

Historically, U.S. immigration law has been weak on labor

---

[119] For more on this, see the running account in *Interpreter Releases*, notably Sam Bernsen, "The Proposed Restriction of the 'B-1 in Lieu of H-1' Concept," September 13, 1993, pp. 1189-1193, and "State Dept. Proposes Restricting B-1 in Lieu of H-1 Visas," July 26, 1993, pp. 968-970.

[120] One of the more intense segments of the previously mentioned "60 Minutes" program came when Leslie Stahl, a "60 Minutes" anchor, was shown walking down a hallway with a Hewlett Packard executive who was refusing to be interviewed.

standards for H-1 nonimmigrants (the old term).[121] Until 1990, INS alone decided whether the wages offered to an alien of "distinguished merit and ability" were appropriate for the desired alien worker. There had been some controversies about the program, primarily regarding nurses and to a lesser extent about show business people and engineers.[122]

Rewriting the immigration law in 1990 involved a great deal of horse-trading, not always well thought out in advance. Special arrangements and newly-lettered visas, with some labor market protection provisions, were made for nurses and for show business workers. Without getting into all the complications, suffice to say that the residual nonimmigrant workers in the H-1B category were subject to much discussion. In the end, the restrictionists won the LCA provision, which at least placed the H-1B wages on the public record, and obtained the 65,000 cap; employers were able to extend the maximum term of these "temporary workers" to six years,[123] and the workers no longer had to argue that they retained a permanent residence in another country.

Nominally, the 1990 Act calls for employers of H-1Bs to pay the prevailing wage, which is supposed to be established by the State Employment Security Agencies (affiliated with the U.S. Employment Service and largely funded by DOL). The over-stretched state agencies have little ability or staff to set these wages and none to enforce them. In effect, employers file the LCAs, creating something that appears plausible in terms of a prevailing wage, and INS approves the petitions. (There is no provision in the law that DOL must decide on the appropriateness of the newly filed LCAs; it simply

---

[121] This has been the case for virtually all of the nonimmigrant visa categories; most of what little labor market regulation there was has been in farmworker programs, such as the bracero program (1943-1964) and the ongoing H-2A program, with DOL playing the major role.

[122] See, for example, D.S. North *Nonimmigrant Workers in the U.S.: Current Trends and Future Implications*, Employment and Training Administration, U.S. Department of Labor, Washington, 1980, pp 127-142. Virtually all of this section of the report deals with foreign-born nurses.

[123] The norm in the past had been for a one-year visa, with the easy possibility of a two-year extension and a more difficult prospect for longer extensions.

must receive them, make them available to the public, and hear protests from interested parties, defined broadly. There have been only a handful of protests, as neither organized labor nor the engineering associations have done much with this opportunity.)

The 1990 Act, in short, does little more to protect adversely affected U.S. workers than did previous legislation. There are, however, some signs of administrative change in this field; these changes are covered later.

The Nonimmigrant Pattern: Numbers, Characteristics

One must work with estimates, and not very satisfactory ones, when dealing with the nonimmigrant pattern in industry.

Numbers. Because of the nature of the system, the data are weaker with the B-1 program than with the H-1B. There is, for example, no recording of occupation in the case of the B-1 visa, only the nationality of the visa holder. Further, it is clear that the vast majority of all B-1 visa holders are *not* engaged in science and engineering work in the U.S., so that 2.7 million admissions figure is not helpful.

The only estimate we have seen for the B-1 program, from its leading critic, is 30,000; this is presumably an estimate of the number of such workers in the U.S. at a given time. The critic, James Schneider, interestingly, is not a worker activist; he is owner of a San Francisco-based consulting firm in the software business, has been hurt by others' use of B-1 workers, and is a spokesman for the National Association of Computer Consultant Businesses (NACCB.)[124] The 30,000 seems a little high, simply on the grounds that the U.S. Census shows a total of 470,000 computer professionals (programmers and others) for 1990. Given that total, the 30,000 estimate is almost certainly exaggerated, but the actual number is probably not trivial, and those involved appear to be concentrated in a particular industry. No alternative estimate seems to exist.

As for the H-1B pattern, there is a plausible outer limit. There

---

[124] The estimate was quoted in the previously cited *San Francisco Chronicle* story on the subject; the reporter said he tried the estimate on people in the business and one or more persons at INS, and those consulted found it plausible.

were an estimated 7,800 computer and mathematical scientists and 17,800 engineers (programmers could be placed in either category) admitted as H-1Bs in 1993.[125] Many, perhaps most of these were singleton admissions, along the lines previously described, but thousands could have been computer programmers admitted through group actions. The estimated number of computer and mathematical scientists arriving in the H-1B category soared from 4,000 in FY '92 to nearly 8,000 in FY'93 — a troublesome set of statistics in view of the "60 Minutes" report.

Characteristics. The programmers seem to be overwhelmingly male and English-speaking. Most of the newspaper accounts discuss India as a prime source of these workers, and Indian "body shops" as the middlemen between the programmers and the U.S. firms.

Great Britain, Australia, and Canada are also frequently mentioned as sources of these workers. Microsoft, for one, is often mentioned as recruiting H-1Bs heavily in Canada.

It was British engineers and designers, some years ago, who were said to be taking engineering and design jobs in the U.S. under the old H-1 program.

The Nonimmigrant Pattern: Sector and Occupation

As to the sector of the economy where the B-1s and group H-1Bs concentrate, virtually all of the anecdotal evidence to date points to the software industry. There appears to be no mention of this pattern in government, in academe, or even in traditional manufacturing (though that would be a likely possibility).

Computer software is a growing industry (as traditional manufacturing is not) and one that uses large numbers of professionals who often work in teams. The nature of the industry is such that it can work fairly easily with either the B-1 or the group H-1B system. The apparent concentration in a single line of work (programming) in a single industry (software design) suggests a more dramatic impact than a diffusion of these workers through many industries and occupations.

---

[125] See both Exhibit Eight and the Appendix.

The Nonimmigrant Pattern: Wages, Working Conditions

There is nothing but anecdotal evidence, previously cited, for the wages paid to the B-1 programmers and probably no way to secure more concrete evidence short of a full-fledged government investigation of this operation.

As to the wages paid to H-1B programmers, we have two journalistic sources. Pete Carey, a reporter for the San Jose *Mercury News* — in the heart of Silicon Valley — told us that his newspaper had devoted substantial resources to the question. Carey flew to Washington, secured a tape from DOL of all the LCA wage data, and enlisted the newspaper's database specialists in an effort to analyze it. They came to the conclusion that no meaningful means,

---

One Programmer's Story

Alex Dubenko...is a Ukrainian computer analyst recruited by a Kiev company to work in the U.S. On arrival here last October he was met by the company's local recruiter, a Belgian, who put Dubenko and several Russians to work at firms in Maryland and Virginia.

Dubenko and three other men were lodged in a house near Manassas [Virginia], driven to and from work, and fed. Dubenko said he was paid $20 a week, which, after conversion to rubles he calculated as a proper wage for equivalent work in the Soviet Union.

Unsure of his legal status, Dubenko worked seven days a week, often morning to night. He was given a raise to $40 a week.

Finally, in frustration, he and one of the other men began calling local companies last February to look for another job. Although neither had a valid driver's license, they took the recruiter's car and found their way around the Beltway, out Interstate 270 to [Fred M.] Shulman's company in Rockville [Maryland].

They applied for work, passed a test and were hired, according to Dubenko. When the recruiter found out, Dubenko said, he ripped the telephone from the wall, removed the battery from the car and said, "In two days you fly to Russia."

Dubenko said he found an extension phone and called Shulman's company, which sent someone to pick them up. Dubenko said he is now earning about $35,000 a year, working under a proper visa.

"But there are still people living in the house in Manassas," he said, still working for a foreign recruiter.

from the *Washington Post*, Nov. 9, 1992

medians, or modes could be secured either by occupation or by region. If you wanted to know about a specific job opening, the wage offer could be examined, but, he said, no effort had been made to preserve the data in such a way that one could study trends and patterns. (Collecting the LCAs is a task assigned to the Employment and Training Administration, not the Bureau of Labor Statistics, which might have had a different approach.)[126]

Carey wrote of one programmer's experience:

Some workers appear to have little choice but to work for the company that brings them to this country. A contract signed by Kala Sivasubramanian, who was hired by Tata Consultancy Services of Bombay, India, and who worked for Oracle Corp. in Redwood City [CA], contains an agreement that employees will not leave Tata to work for another company, according to court records.

If Sivasubramanian did, her contract states, she would be liable for $30,000 in damages. She left, and Tata is suing her. Under the contract, the Bombay programmer surrendered her passport to her bosses for 'safekeeping' was discouraged from bringing her family or from getting pregnant...[127]

The contribution by "60 Minutes" to the question of wages and working conditions was a TV screen full of LCA forms showing $26,500 a year, which one or more H-1B employers had decided was the prevailing wage for programmers. The actual prevailing wage is considerably higher (but this has not been established in any meaningful way).

We now turn to the apparent impact of foreign-born S/Es generally on various U.S. institutions and populations.

---

[126] Several telephone conversations with Carey in the spring of 1993.

[127] The *Mercury News* ran a major series of articles on immigration from May 29 to June 3, 1993; the quotation is from one of them: Pete Carey and Steve Johnson, "Abuse mars skilled-worker program," on the paper's front page, June 3, 1993.

Chapter Five

Impacts of Foreign-Born Scientists and Engineers

The presence in the U.S. of large numbers of talented and hard-working foreign-born scientists and engineers has been very soothing to the American Establishment. It has prevented the establishment from having to pay serious attention to a number of painful issues, such as the distribution of income between scientific and managerial talent, and between science and engineering workers on one hand, and stockholders and taxpayers on the other. The presence of this talent has dampened the pressure to make major reforms in K-12 science and engineering education and has eased, if not eliminated, the pressure to recruit American women and American Blacks for science and engineering careers. (Although we have little hard supporting evidence, our suspicion is that white American males are more comfortable working with male foreign-born S/Es than they are with native-born women or Blacks, who are frequently more assertive about their rights than the foreign born.)

Further, the presence of foreign-born S/Es has helped academe not only preserve but also expand graduate educational institutions, and, as we will argue, maintain high overhead rates for their government grants. Similarly, the presence has helped U.S. industry maintain progress in research and development — all without wrenching side effects.

All of these comments are addressed to what we previously defined as the *immigrant* pattern of foreign-born S/Es; what we see are a series of indirect, often mixed impacts made by a number of people moving one at a time through the systems described. The impact of the *nonimmigrant* pattern, seen so far largely in the software industry, is much more dramatic and much more exploitative than that of the immigrant pattern.

In the balance of this chapter, we explore the impact of foreign-born S/Es on education, on industry, on U.S. populations, and on science and engineering *per se*. These comments are, of necessity, impressionistic, but so are all statements about the impact of a complex phenomenon on a series of complex systems.

The Impact on Education

The presence of foreign-born S/Es has a number of different impacts on different segments of U.S. education. In this section, each segment is discussed separately.

The Impact on K-12 Education

There are foreign-born youngsters growing up in the U.S. and taking chemistry, biology, and physics classes in high school. Many of them, particularly among the Asians, are doing well, winning Westinghouse Science awards and similar contests. But this is part of the mainstream of American life, not part of the deliberate, adult migration stream we have been describing. Thus, it is beyond the scope of this report.

If anything, the migration of adult foreign-born S/Es has dulled interest in improving the not-very-impressive record of America's public schools in science and mathematics. Were the nation to be experiencing a shortage of scientists and engineers — as opposed to seeing many of these graduate slots and jobs filled by talented foreigners — a greater effort would presumably be made to improve this part of America's educational system.

Further, in contrast to the extensive use of foreign-born physicians and nurses in the nation's big-city public hospitals, there has been little visible employment of foreign-born S/Es as teachers of science and mathematics in the public schools.

> **An Exception**
>
> As a participant-observer, the author was waiting to take the Graduate Record Examination at Seton Hall University. He decided to strike up what might be a useful conversation about foreign graduate students in the U.S.
>
> There were scores of young Americans in their twenties, not very useful in this context, but there was one exception. He was the same age as the others, but was the only person wearing a suit, a somewhat worn, drab black suit, with a white shirt and a somber tie. He had very dark skin and was most likely from Africa.
>
> It turned out that he was Nigerian, holding both BS and MS degrees in mechanical engineering, and was interested in getting a PhD. Where had he gone to university? Moscow. Did he have a student visa (F or J)? There was no need for that; he had received an immigrant visa (given his degrees) easily from the U.S. Embassy. Before asking the next question, the author thought about his multiple disadvantages in the U.S. job market: Black, alien, shy, and with a Russian technical education.
>
> What was he doing for a living? He was teaching remedial math at a community college in the center of Newark, N.J., and was not very happy doing it.
>
> He was the only foreign-born S/E the author met in two years who was even close to teaching in the K-12 levels.

The Impact on Undergraduate Education

As a group, foreign-born S/Es make only a mild direct impact on undergraduate education; as noted earlier, they are only a small portion of the undergraduate population, even in science and engineering studies.

They play more important roles as not-very-well-paid TAs, and to the extent that they fill TA jobs that would otherwise be filled by the faculty, they help the universities with their low-wage labor. Increasingly, they are moving into faculty positions, particularly junior ones. There is, as noted earlier, some controversy about their role as TAs and faculty members, on the dual grounds that their ability with English is sometimes marginal and that many of them carry with them cultural baggage regarding the appropriate roles for women and members of America's resident minorities.[128]

---

[128] George Borjas' findings on this point are discussed in footnote 39.

The Impact on Graduate Education

It is at the graduate level that foreign-born S/Es make their greatest contribution to, and impact on, American education. Foreign-born graduate students, it is generally agreed, are helping maintain the strength of a major American intellectual structure, graduate education in science and engineering.

In recent years foreign-born S/Es have kept graduate departments from getting smaller by providing the needed enrollment. In addition, by doing much of the day-to-day research that maintains the flow of grants, they have helped keep research dollars rolling in. These are major impacts, widely enough recognized by others, so that there is no need for further comment here. What is not so well known is that:

1. These benefits to science and engineering graduate education have come without any major, badly-needed, but painful, adjustments on the part of the universities, or society at large;

2. The flow of science and engineering grad students not only has prevented a contraction of graduate departments, despite the lower USC interest in the departments' offerings, it has helped them grow remarkably in recent years;

3. The availability of low-paid, foreign-born research assistants (RAs and postdocs) has helped, we argue, the bigger research universities move government moneys for research into university overhead accounts; and

4. The presence of foreign-born applicants for graduate school has led to some displacement of native-born citizens seeking those slots.

Let us examine these four points in more detail.

No Major Adjustments. The large number of talented foreign-born science and engineering graduate students has allowed the status quo to continue virtually unruffled in the graduate schools. Because of their presence, there was no need to cut back programs because of the reduced demand from native-born students; there was no need to press strongly to recruit (although some of this is done) from among nontraditional potential American sources of science and engineering graduate students, such as women or resident minorities; and there was no need to make significant increases in graduate school stipends or postdocs' salaries[129].

Similarly, the graduate schools were kept busy without need for the broader society either to increase science and engineering incomes generally, vis-a-vis those in medicine and law, or to increase incomes, more specifically, for PhDs in science or engineering.

Expansion of Graduate Departments. Most writers, when discussing the interaction between foreign-born PhD students and science and engineering graduate departments, speak of the presence of foreign-born students as preventing a contraction in these programs. What they rarely mention is the role that the foreign born must have played in the last couple of decades in the expansion of the number of graduate programs in science and engineering, often by the creation of very small programs, which Bowen and Rudenstine view with understandable (if muted) concern.

Two sets of numbers are instructive. There were 130 graduate programs granting PhDs in Physics and Mathematics in 1958; by 1988, there were 282 of them, an increase of 117%.[130]

But much of this growth was in what the two university presidents regard as either small or very small PhD programs; they defined the very small ones as those that produced fewer than two degrees per year on average in the 1980s, and the small ones (which includes the very small ones) as those that produced fewer than four degrees on average. The numbers of programs meeting this description grew as

---

[129] Noted earlier was the pattern of the NIH funding of postdocs, which barely kept up with the level of inflation. Similarly, Finn makes this observation about stipends: "Interestingly, while U.S. universities truly want to recruit talented U.S. graduate students, they do not seem to have raised their research assistantship and teaching assistantship stipends as much as they have raised faculty salaries in recent years. Had they done so, there would be more U.S. students competing for places in the Ph.D. programs." See: Michael G. Finn "The Stay Rate of Foreign Students Who Earn Science and Engineering Degrees from U.S. Universities," a paper presented to the Pacific Science Association, Honolulu, May 27-30, 1991, p. 6.

[130] Bowen and Rudenstine, *op. cit.*, Table G. 4-1.

follows:[131]

Small Departments	1958	1988	Increase
Physics	13	75	477%
Mathematics	13	74	469%

The number of very small departments, in these two disciplines, grew slightly more rapidly.

Of greater concern was the growth in the number of degrees provided by departments that were placed (by their academic peers) in the lowest (or fourth) tier of quality. In 1958 there were 463 PhDs awarded in mathematics and physics to students in first tier institutions; the number had grown to 1,011 in 1988, an increase of 118%. But fourth-tier degrees in the same fields grew from 18 to 235 in the same period, an increase of 1,206%.[132]

While Bowen and Rudenstine did not deal with citizenship in their book, we know from other sources that the smaller and less prestigious institutions have the highest percentage of foreign-born graduate students (at least in science and engineering). It would appear that the expanding number of foreign-born graduate students helped expand the number of science and engineering graduate departments, including many small ones, and many in the bottom-quality tier. Was that a good idea?

University Overhead Rates. As an associate of several nonprofit research organizations and as an occasional member of panels reviewing funding proposals to government agencies, the author has been exposed to the subject of overhead rates fairly continuously for a quarter of a century.

The overhead rates of academic institutions and other nonprofit organizations are often cloaked in complex accounting formulae and regulations, but in the end they, like profits in a for-profit context, are what the market will bear. Further, as in much of life, those who have, get; overhead rates are reimbursements for expenditures made

---

[131] *Ibid*, Table G. 4-4.

[132] *Ibid*, Table G. 4-2.

that subsequently are permitted to be classified as overhead. Institutions with little money to work with cannot make the initial expenditures (hiring deans, providing staff training, and the like) to secure high overhead rates in the future.

Some clients — and the U.S. government is usually quite relaxed about this — will accept fairly high overhead rates that are based on the contractor's own history, but most clients have a limit as to how much money they will pay for a specific activity. If a contractor with a plausible proposal is bidding on a project on which the client will spend $100,000, it does not matter very much to the client if the overhead rate is 100% or 50%, just so long as the total is no more than $100,000, the budget seems believable, and the rates meet the formal requirements of the accounting system.

Thus, in a very general sense and over time, the contractor can win with low direct cost/high overhead bids, or with high direct cost/low overhead bids, but, in competitive situations, not often with both high direct cost *and* high overhead bids. Over time, truly major contractors, like the research universities, have created traditions of high overhead operations; and, over time, much government work is let under noncompetitive conditions.

Meanwhile, the low salaries of the RAs and postdocs, who are often foreign born, help perpetuate the universities' continuing ability to secure high overhead rates on government scientific and engineering research contracts; we have not seen this linkage suggested before, but it seems an obvious one.

Some attention has been paid by both the media and Congress to high overhead rates and flat-out waste in some government contracts with major universities.[133] One reads, in one of those blue-bound, carefully-written, nonflamboyant reports of the General Accounting Office:

> ...[W]e found almost $1 million in unallowable...costs charged to the government by Stanford. Of this amount, about $184,000 was paid as the government's share of the depreciation of a 72-foot luxury yacht...In another

---

[133] Here there are two different problems, which are not always distinguished; there are clearly some inappropriate items among the overhead expenditures as the GAO report (see following footnote) suggests; there are also high, but legitimate overhead expenditures by some universities. Bringing the latter under control is a budgetary problem, not a law-and-order problem.

case, MIT identified $778,000 in overcharges to the government...[t]hese overcharges included the costs of such items as floral designs, dues for airline airport clubs, art work, overseas trips, receptions...Harvard Medical and Berkeley both charged other unallowable or questionable items to the government...Charges related to the costs of operating the residences of university administrators occurred at all four universities we reviewed.[134]

Perhaps it is only a coincidence, but Harvard is the nation's leading employer of foreign-born postdocs with 1,065; Berkeley is fourth, with 378; Stanford fifth, with 303; and MIT nineteenth, with 170.[135]

Clients *can* do something about overhead rates, if they insist. A near-at-hand example is this study and others funded by the Alfred P. Sloan Foundation. Sloan will not accept an overhead rate of more than a modest 15% and makes no grants to individuals, so researchers must be associated with an institution willing to accept that rate. Sloan's funding of research has not been brought to a halt by its position on overhead rates, and one suspects that the government could continue to fund research with more modest overhead rates than it now pays.

Sloan, however, has more freedom than the federal government does in this matter. The Government would need a major policy change — possibly including legislation — to bring overhead rates down to the Sloan level. Were it to do so, it might find that some of the major research universities would simply refuse, or threaten to refuse, to accept the grants.

The basic point remains that the low wages of foreign-born RAs and postdocs makes it easier for universities to maintain high overhead rates.

Some Displacement. Although graduate schools favor USC applicants, there are enough very able noncitizen applicants to displace some of the U.S. citizen applicants. This is apparently not

---

[134] U.S. General Accounting Office, *Federal Research: System for Reimbursing Universities' Indirect Costs Should Be Reevaluated*, (GAO/RCED-92-203), Washington, D.C., 1992, p. 17.

[135] J.G. Huckenpöhler, *Foreign Participation in U.S. Academic Science: 1991*, National Science Foundation, (NSF 93-302) Washington, D.C., 1993, Table A-26.

a one-to-one phenomenon; Finn estimates that the displacement effect is no higher than 25% — for every four foreign-born applicants accepted, one native-born applicant is rejected. Assuming both the institutional disposition to accept USC applicants and the relatively low volume of qualified USC applicants (given the current incentive structure), we see no reason to doubt this estimate.[136]

The Impact on Industry

As with academe, the foreign-born S/Es have a soothing impact on American industry (and government as well). They are a talented work force that is obtained without any strain on management systems, and there are enough of them when combined with the remaining native-born S/Es to prevent wages from rising much faster than inflation. It appears that the foreign-born S/Es are most useful to U.S. industry if they stay in the U.S., as most of them do (which we will call Scenario One), but they are also useful if they return to their home countries (in Scenario Two).

Scenario One: The Foreign-Born S/E Stays in the U.S.

Industry benefits from the foreign-born S/Es whether they move through the immigrant or the nonimmigrant patterns mentioned earlier.
The Immigrant Pattern. The principal benefit to industry is the ready availability of well-trained, bright, hard-working foreign-born scientists and engineers to work in industry's laboratories and factories. They meet a major portion of industry's needs at the PhD level and play significant roles at the MS level as well.

Further, they are an attractive work force to American management, just as they are to academic management. As has been previously mentioned, they often come from hierarchial societies, and most have spent years in one of the most hierarchial segments of American life, academe, so they fit neatly into the corporate world.

---

[136] Finn, "Foreign Engineers in the U.S. Labor Force," in *Foreign and Foreign-Born Engineers in the U.S., op. cit.*, pp. 100-102. Finn, in commenting on an early draft of this report, wrote at this point that this is "...a subject that has not been studied much and my estimate was based on the best information available and should not be taken as definitive."

> **Vacation Plans**
>
> He was an electrical engineer from Pakistan, and in addition to a BS from the University of Sindh, he held an MS from a Canadian university. He was explaining at an IEEE meeting why he thought foreign-born engineers do well in America.
>
> "I had vacation plans, tickets, everything; my boss told me two days before I planned to leave that he had a major, unexpected crisis and he needed me to work on it. I junked my vacation for him. Americans don't do that."

By American standards, most are relatively docile workers, accustomed to doing what their superiors tell them to do. In addition, as G. Pascal Zachary of the *Wall Street Journal*, suggested to us, they bring into corporate America an element of the Third World that is useful to management — their attitude towards their families:

"If they feel that they have to work long hours they simply tell their wives that this is the way it is; they are considerably less likely than American men to let family obligations interfere with demands on the job."[137]

In addition, if the foreign-born S/Es were not around or were leaving the country in large numbers, it is likely that there would be greater upward pressures on salary levels than there currently is, particularly at the PhD level, where fewer Americans are available than formerly. As it is, industry is relieved of any major pressure to try to shift financial reward patterns from law and medicine, on the one hand, to science and engineering on the other. Similarly, industry is relieved of having to increase the differential pay for PhDs vis-a-vis those with lesser degrees because of the availability of the foreign-born PhDs.

The presence of the foreign-born science and engineering talent has a similarly soothing impact on the federal government. Absent this work force, the government would have to press harder for educational reform at the K-16 levels and would have to push for greater recruitment of women and resident minorities into science and engineering.

---

[137] Conversation with G. Pascal Zachary at IEEE meeting in Portland, Oregon, September 4, 1993.

Speaking of the federal government, it is probably the least adept user of this resource — foreign-born S/Es — of any major sector of the economy. Looking at the skill with which the various sectors approach this resource, we would (again these are impressions) rank them as follows: academe, industry, local government, federal government.

Academe knows more about how to recruit, hire, and utilize foreign-born science and engineering talent than any other institution in the nation. Foreign student advisors and their colleagues[138] know their part of the immigration law backwards and forwards and work on these matters on a full-time basis. Given their level of expertise and the differential salary patterns of university administrators and corporate lawyers, they probably can handle a specific visa matter for a small fraction of the cost that a corporation would have to pay to solve the same problem.

The impact of foreign-born S/Es on industry would no doubt be greater were industry as knowledgeable as academe; we have already noted the fact that at least some and perhaps many companies do not hire foreign-born S/Es unless they already have green cards — a posture that *no* college or university would take. Further, there is other evidence of the lingering corporate lack of knowledge about how to handle the immigration law in the employment setting. This was contained in a National Research Council publication on foreign-born engineers:

> As far as foreign nationals are concerned, the Immigration and Nationality Act provides that a foreign national (already in the United States) who seeks immigrant status for the purpose of employment is ineligible to receive such a visa and is then excluded from admission in the United States at this time.[139]

---

[138] There are specialties within specialties. The author spent an informative hour talking with the executive at MIT who headed an office dealing with the foreign-born; she did *not* work with students (graduate or undergraduate), *nor* did she work with the faculty. She and her staff did nothing but cope with an intermediate class, foreign-born scholars and nonfaculty researchers.

[139] Peter Cannon, Rockwell International Corporation, "Foreign Engineers in U.S. Industry: An Exploratory Assessment," in *Foreign and Foreign-Born Engineers in the United States, op. cit.*, p. 111.

It is clear that neither the author nor the editors of this publication were aware of: (a) labor certifications, (b) the option of adjusting from nonimmigrant to immigrant status within the U.S., (c) the difference between exclusion and deportation (admittedly a fine point), or (d) the general ease with which reasonably informed employers of alien S/Es handle such matters.

The federal awkwardness with this population relates as much to civil service rules regarding the employment of PRAs and nonimmigrants as anything else.

The Nonimmigrant Pattern. The nonimmigrant pattern of hiring groups of B-1s and H-1Bs, particularly in the software industry, brings immediate benefits to industry, in the sense that they pay considerably less for workers than they would otherwise.

Even if these nonimmigrant programs are reduced or eliminated (in the case of B-1) or made a little more expensive (in the case of H-1B), the fundamental benefits of the immigrant program, described above, are likely to continue for years to come.

Scenario Two: The U.S.-Trained Foreign-Born S/E Goes Home

Politically, it has often been useful for the United States that many ranking politicians from other nations have attended college or graduate school in the United States. They often had a good experience here and picked up various ideas that were important to them, and sometimes to the U.S., in later years.

Similarly, there has been speculation that the foreign-born S/Es who do, in fact, go home after their education in the U.S. (probably a minority of foreign-born S/Es) often nonetheless provide benefits to U.S. industry in the years that follow.

On the narrowest of levels, in the U.S. the S/Es learned to work with particular hardware or software and then ask their employers to purchase it for use in the homeland. More broadly, they become linked to various networks that include U.S.-based individuals, institutions, and corporations, and they work with them in later years. In some cases this evolves into cooperative ventures between homeland and U.S.-based corporations.

There is a potential downside as well to the departure of U.S.-trained S/Es: They may take technical knowledge, secured in the U.S., probably at the expense of U.S. taxpayers, and use it in competitive situations to deny U.S. firms market share. Presumably this is an ongoing process, and to some extent Asian S/Es, notably in Japan and in the Little Tiger nations, are engaged in exactly that activity. This is occasionally mentioned in business circles.

What is less frequently mentioned, since the end of the Cold War, is the departure of U.S.-trained S/Es to work for hostile powers. Clearly, Iraq was on the edge of building a remarkable cannon, thanks to departed North American expertise. And, on the other side of the coin, Nazi Germany lost a tremendous asset when it caused so many of its scientists to leave before and during World War II.

The Impact on U.S. Populations

It is always difficult to work out the economic impact of a migratory population on a stationary population. What would have happened to the resident population if there had been no migrants? What would have happened to the resident population if there had been twice as many migrants? half as many? What if the timing of migration had been different? What labor market adjustments would have been made if the migrants had not been there?

One theory, which makes more sense in science and engineering than in many other segments of the economy, is the key-worker notion. It is often argued that many resident workers in complementary jobs would have lost their livelihood had it not been for the timely arrival of, say, the foreign-born engineers who design the autos that resident workers manufactured and serviced? Some of that may be operating in science and engineering in the U.S. today, but the argument gives little credit to human ingenuity and the multiple ways to cope with what appears to be a shortage of skilled labor (e.g., accelerated training of natives, higher wages, redesign of jobs, longer hours for the skilled natives, the tapping of previously untapped personnel resources, re-hiring the recently-retired, etc.).

Moving away from that set of intangibles, it is probably fair to say that the foreign-born S/Es who attended graduate school in the U.S. and remained in the U.S. work force had these two generally positive impacts on U.S. populations:

1. They raised the average skill level of the U.S. labor force generally;

2. They increased the size of the science and engineering professional labor force, (thereby presumably leading to some increases in production).

But they also had these five generally negative impacts:

1. For S/Es, generally, decreased pressures to increase wages in the field;

2. For S/Es, generally, loosened the labor force, leading indirectly to a little more unemployment;

3. For some middle-aged native-born S/Es, hastened their retirements;

4. For women, dampened efforts to increase their opportunities in the field; and

5. Similarly, for Blacks, Hispanics, and Native Americans, slowed efforts to increase their opportunities.

This then, is one way of characterizing the impacts of the foreign-born S/Es moving through the previously-described immigrant pattern of foreign-born utilization; it is clearly a mixed bag. (The negative impacts of the nonimmigrant pattern are rather more dramatic, and less mixed, than the picture shown above.)

Since the positive impacts have been described fully by others, let us focus at least briefly on the five more negative consequences.

Wages

While different observers place different significance on this matter, there seems to be little disagreement that the presence of so many foreign-born S/Es has depressed the science and engineering labor market. This is a passage from two writers who are very enthusiastic about foreign-born engineers in the U.S. labor market:

> We must acknowledge, as the did the Committee on the International Exchange and Movement of Engineers, that by their very presence, foreign-born engineers probably depressed earnings below what they would

otherwise be. If foreign-born engineers had been barred from the U.S. workforce in the 1980s, there would most likely have been an increase in engineering salaries above then-current levels, particularly for those American engineers with the PhD. In turn this could have sparked an increase in PhD enrollment.[140]

But wages are not simply a matter of supply and demand; fashion plays something of a role, and in recent years science and engineering have not been very fashionable. Further, there is the power of a specific work force to wrest economic rewards from the balance of society; as we show later, S/Es (unlike physicians and lawyers) have been remarkably inept in this regard.

Unemployment

Science and engineering unemployment rates are always much lower than for the labor force as a whole, but in recent years they have grown rapidly, from an admittedly small base.

Unpublished BLS data show the following, with the '93 data being for the first six months of the year:[141]

Discipline	Unemployment 1990	Unemployment 1993
Natural Science	1.5%	2.4%
Mathematics and Computer Science	1.5%	3.1%
Engineering	2.1%	4.2%

Similarly, and using a much larger survey base, the annual American Chemical Society's (ACS) salary survey reported that unemployment among chemists had increased from 1.1% in March 1990 to 2.0% in March, 1993.

The industry trade paper, *Chemical and Engineering News*, suggested a chemist's equivalent to the meteorologist's discomfort index (a single number covering both temperature and humidity); this

---

[140] Endelman and Loughran, *op. cit.*, p. 15.

[141] The fax from BLS identified the source as Labstat Series Report.

would be:

> ...a total of the percentages of chemists who are unemployed but looking for work, employed part time, or in postdoctoral or fellowship positions. In good economic times this hovers close to the 4.5% mark. As of March 1 of this year it was at 7.2%, the highest it has been in 20 years.[142]

S/Es were then suffering higher than normal unemployment rates for a variety of reasons: a lingering recession, an effort by major firms to downsize, a reduction in corporate research and development, and, affecting engineers particularly, the reduction in U.S. defense expenditures.

All of this was happening at a time, as was shown in Exhibit One, when international movements of S/Es continue to rise. Clearly there is not a one-to-one relationship between the admission of one foreign-born chemist and the lay-off of another one (foreign born or native born), but loosening labor markets take their toll, and the timing of the two trends could not have been worse.

Earlier Retirement

The nonvoluntary, earlier retirement argument goes like this: corporations, reacting to defense cutbacks and recessionary pressures, are forcing early retirements on scientists and engineers who would rather work for a few more years. At the same time newly graduated foreign-born S/Es are hired, often in the same corporations.

Thus, there is a struggle not only between the old and the young but between the foreign born and the native born (who constitute a high proportion of the older S/Es). This is clearly a subset of the general unemployment question discussed above.

Let's assume that a single corporation is following this policy. There is a certain rationale behind it; young people are less expensive to hire than older ones are to retain; young S/Es have just experienced the latest in technical training (which the corporation may not have made available to its older workers); young people (perhaps especially the foreign born, who have few roots in the U.S.) will relocate more easily than older ones.

---

[142] Michael Heylin, "Job Market for Chemists Remains Depressed, Salaries Gain 5%," *Chemical and Engineering News*, July 12, 1993.

It is not just a preference for youth or just a preference for the foreign born, but it's hard to convince the 55-year-old native-born engineer who has worked for the company for 30 years that what is happening to him is simply the outcome of a rational personnel policy. He places a higher value on in-company experience and on downward loyalty than the employer does, but his letter of complaint is more likely to go to his Congressman than to a member of the employer's board of directors.

Women

Every woman (as well as many men) with whom we discussed foreign-born S/Es recognized a tension between the presence of large numbers of talented foreign-born S/Es and equal opportunity for women in science and engineering. This was particularly true for women *working in academe*, even those whose full-time job was to facilitate the presence of foreign-born S/Es at the universities.

Their arguments contained three threads, which they discussed with differing levels of intensity. There was the reasonably neutral matter of numbers; if there were not so many talented foreign born in graduate schools and in industry, universities and corporations would be pressed harder to recruit more women, Blacks, and Hispanics for science and engineering. (The foreign-born S/Es, as we noted earlier, include a lower percentage of women than the native-born S/Es.)

Then there was the matter of what the women regarded as American society's sexist and racist attitude toward mathematics and science; women and girls (and Blacks and Hispanics) are not expected to handle such matters. As a result, in all sorts of direct and indirect ways, they are encouraged to pursue other interests. Most of the women had personal stories, usually about a high school guidance counsellor steering girls away from mathematics or science. An attitude of understandable and unmixed indignation prevailed.

Finally, there was the question of how foreign-born S/Es, in their roles as TAs and faculty members, treat women and minority students. The general impression was that many of them, carrying homeland views on gender and race, did little to encourage women and minorities in these studies, and that, since the TAs played a major role in undergraduate education, many women and minority students as a result were lost to science and mathematics.

It was painful for these college administrators to say these things because they appeared to be liberals to the core, and good liberals are reluctant to fault a foreign-born population. It was particularly hard for them to argue, as they generally did, that the foreign-born TAs and faculty members were *worse* on these matters than the native-born white males who usually run these institutions.

Our sense is that the all-too-common foreign-born attitude on gender and race is a major problem and one that is much harder to solve than the question of the language abilities of the same TAS and faculty members.[143]

Despite these influences, however, women are securing a growing percentage of science and engineering PhDs:[144]

Discipline	Percentage of PhDs that are Women		
	1962	1977	1992
Physical Sc.	4%	10%	20%
Engineering	.3%	3%	9%
Life Sciences	10%	21%	25%

But compare the above percentages of women in a graduating class to these for 1989: dentistry, 26%; medicine, 33%; and law, 41%.[145] These fields, traditionally also male-dominated, are now apparently considerably *more* accessible to American women than science and engineering. Do the higher financial rewards of these professions (than S/E) attract more American women? Is it the lack of competition with foreign-born males for these graduate slots that opens up more opportunities for American women? We wish someone would look into these questions.

While discussing this subject, we ran into an anomaly, which may

---

[143] It is also a problem that is routinely denied by American liberals, whether they be scholars or government officials. The author once spent some energy trying to build coalitions among Southeast Asian and Black organizations, notably in Atlanta, getting a warm reception from the Blacks and a cold shoulder from refugees, who had treated the author well on other issues.

[144] *Summary Report 1992, op. cit.*, p. 19.

[145] *Stat. Abs. U.S. '92*, p. 174.

be of interest but is probably of little statistical significance. One of the reasons that some foreign-born S/Es stay in the U.S., we were told, is because of heavy pressure from their wives to do so. The wives, particularly those from Japan and Latin America, it is said, dread the prospect of returning to their homelands, because they will have considerably less freedom there than in the U.S. So, to the extent that women's rights keeps foreign-born male S/Es in this country, it tends to increase (or at least stabilize) the percentage of males in the S/E labor force.

An Ideological Digression

Before discussing the impact of foreign-born S/Es on potential S/Es among the nation's resident minorities, it may be useful to digress for a moment about why Black and female and Hispanic participation in S/E matters.

It essential that all members of society be represented in positions of power, in government, in industry, in law, and in medicine, so that policies can be made that are fair to all, and so that policies are implemented with an understanding of the separate subcultures in which all of us live. Multi-racial legislatures, or groups of physicians, or bureaucracies work best in multi-racial societies such as ours, and it is important that skillful people of all kinds have a chance for leadership roles.

It is extremely important — almost as urgent as it is in the political process — that all occupations be open to everyone, and that people with potential skills as artists, managers, farmers, or scientists have a full opportunity to go as far as they can in their chosen fields. Straight white males should not be discouraged from the dance because of their background, nor should Blacks or Hispanics be discouraged from mathematics or physics. We will be a stronger society, with a stronger economy, if we bring out the highest skills in everyone.

It is also a good idea that we have more people, many more than we have now, who are skilled in science and mathematics; we, as a nation are sadly lacking in these skills, as we are with languages. We need to work harder in our schools, to be a more skilled, and more educated people — and so the rest of the world will not eat our lunch.

Since that is the case, we should reach out to young potential American scientists, engineers, and mathematicians of both sexes and of all colors and faiths. Only if we encourage a broad range of people to compete in these fields will the best of the nation emerge to work in them as professionals for the rest of their lives. Without broad participation in science and engineering we will lose potential talent, and be a lesser society as a result.

Meanwhile, to return to day-to-day matters, we know we need the talent but we can not discipline ourselves to produce it domestically, so we rent and borrow the talent from elsewhere in the world. Maybe that will work, maybe not. Rome, for a while, found it useful to hire foreign soldiers rather than use her own. Maybe the status quo is appropriate in these matters, with a strong emphasis on full male-female, Black-white-Hispanic participation in legislatures, cabinets, and courtrooms, but not in science and engineering. Maybe it is appropriate that most jobs be filled by the Great Ethnic Central Casting Agency, with executives and stockbrokers being Anglo males, with fruit and vegetable workers being Mexican Nationals and Central Americans, with clerical work being done largely by Black women, and with science and engineering becoming the province of the Indians and the Chinese.

The author's preference would be to stir these populations and these occupations thoroughly and to spend some public and private money to see to it that we get a good mix throughout the economy. Further, we Americans should do our own hard work, be it picking apples or devising complex formulae. With those thoughts in mind, let us turn to foreign-born S/Es and America's resident minorities.

Resident Minorities

Does the presence of large numbers of foreign-born S/Es limit opportunities for resident Blacks, Hispanics, and Native Americans? Again a complex and messy question — one involving competition among nonwhites for positions in what had traditionally been a white man's game, U.S. science and engineering.

Perhaps this is answered best if it is answered in pieces; there appear to be at least four separate questions here:

1. Does the presence of foreign-born S/Es tend to dampen the enthusiasm for recruiting domestic minorities for this work (and for improving K-12 instruction in mathematics and science)? Probably.

2. Are foreign-born S/Es using affirmative action slots designed for members of resident minorities? No.

3. Are qualified resident minority graduate school applicants being displaced by foreign-born candidates? In the short run, no, certainly not many.

4. Are financial resources given more readily to foreign-born graduate students than to American Blacks? Yes.

Recruiting Efforts. Tight labor markets have historically opened previously closed job opportunities for various populations; the employment of women and Blacks in the World War II shipyards (previously a sector dominated by white males) is a classic example.

Science and engineering is not a tight labor market and will not be as long as we have a continuing flow of foreign-born S/Es, most of whom stay in the nation. Only if the continuing (mild) pressures to recruit nontraditional workers into U.S. science and engineering for equal opportunity reasons is linked with a much tighter labor market will the nation really do something in this field. So, as suggested above vis-a-vis women in science and engineering, the presence of a full palette of foreign-born S/Es in the labor market vitiates institutional enthusiasm for reaching out to Blacks and other resident minorities. (This is, however, not to deny either the existence or the potential utility of these recruiting programs.)

The concept that foreign-born S/Es may be shouldering native-born Blacks, particularly Black males, out of graduate education has been raised forcefully by Frank L. Morris, Sr., Dean of Graduate Studies at Morgan State University.[146] Dr. Morris points out the grim statistics regarding Blacks in science and engineering graduate schools and lists a whole series of academic specialties (such as civil engineering, oceanography, analytical chemistry, and geology) in which there were numerous doctorates awarded to the foreign born, but not a single one to an American Black in 1990.

---

[146] Frank L. Morris, Sr., "The New Slavery: The Denial of Doctoral Opportunities for African American Students in American Universities in a De Facto Low Wage American Economy," a paper presented to the National Institute of Standards and Technology Colloquium, Gaithersburg, Maryland, Sept. 11, 1992.

He is critical of the trend in federal graduate education support, which is away from granting fellowships to individuals and towards granting research moneys to faculty (who, in turn, hire graduate students as RAs). His point is that direct individual grants can be made only to USCs and PRAs, while research funds can (and do) find their way into the hands of nonimmigrants. He writes, in the abstract to the cited paper:

> The disproportionate use of federal research funding to provide better higher educational access to non American citizens over American minorities especially in the sciences, is an especially tragic and counterproductive example of actions against our collective American interest based upon the low wage strategy and long standing American prejudices against African Americans.

"Slavery," as in the title of Dr. Morris' paper, may be a strong term to apply to the lack of Black graduate opportunities in the U.S., but Dr. Morris is making a valid point. The U.S. should, for reasons previously outlined, make more of an effort to persuade a broader range of our own citizens to seek science and engineering careers.

Misuse of Affirmative Action. To what extent are affirmative action opportunities in science and engineering, designed for resident minorities, being used by foreign-born S/Es? This is a question that has been muttered, if not raised openly, in some restrictionist immigration quarters.

In the first place, there does not seem to be much affirmative action of any kind in graduate schools, other than a mild preference for USC applicants (of all colors). Admission to, retention in, and receipt of a degree from graduate schools seem to be on a nearly total merit basis, so there did not appear to be any graduate school affirmative action program misdirected to foreign-born graduate students.

We did encounter one graduate department that was particularly pleased with its progress on hiring women and minorities into faculty positions (where affirmative action does exist). The department had managed to hire one of the few Black women around with a PhD in her field; it turned out that she was Jamaican, but should she not have been hired for that reason?

We have less information about how affirmative action programs are mounted in industry, but we looked for, and found no reports of the substitution of foreign-born hires for resident minorities in affirmative action settings.

Discrimination in Admissions. Are qualified Black and Hispanic PhD candidates being turned away from U.S. graduate schools because of foreign-born competition?[147]

This does not appear to be the case. Our sense is that under most circumstances, a qualified USC Black seeking admission to a science or engineering graduate school is going to be looked at very, very carefully, and given every opportunity to be admitted.

One problem is that "qualified" and "not qualified" are not hard and fast categories, immediately recognizable to all. There is a continuum of candidates for admission to graduate school, ranging from a few who could get full funding from any institution in the country, to others who would not be accepted under any circumstance by the lowliest tier-four institution. Between these small groups of outliers, there are many very good candidates, many good ones, and many mediocre ones.

In a competitive situation, one's position on the continuum depends on who else is applying and how good they are (or appear to be). The regular input of large numbers of highly qualified foreign-born applicants would tend to reduce the chances of an average native-born applicant. That the average GRE score (see Exhibit Six) of American minority members is well below that of foreign-born applicants complicates the situation.

The other major problem in this field is not the quality of the minority applicants but the quantity. In the U.S., there are not that many minority group members, nor for that matter that many majority group members, who focus on science and engineering early, who study it at college and do well at it, and then, given the incentive structure currently available, opt to study at the graduate level. Merely reaching out during the graduate admissions process is not enough; such efforts to encourage interest in science and engineering must start well before the individuals are undergraduates. (It would help, however, as Morris suggests, if the science and engineering graduate schools did reach out more to the historically Black campuses just before the grad school admissions season.)

---

[147] See also the discussion in Chapter 3.

Financial Resources. It does appear, based on the NRC surveys of new science and engineering doctorates, that Black science and engineering graduate students are more likely to rely on their own resources to finance their studies than the foreign born, and that those who do secure a doctorate are considerably less likely to be debt-free on that happy day than their foreign-born brethren.[148] (See Exhibit Eleven and the accompanying text.)

The Impact of the Foreign Born on Science and Engineering, *Per Se*

Whatever the cost to other U.S. populations, and whatever the benefit to U.S. academic, business, and governmental organizations, the contact between the foreign-born S/Es and America's graduate schools is helpful to science and engineering *per se*.

Most of the foreign-born S/Es currently present in the U.S. were the best and the brightest from their own country, or they would not have gotten past America's science and engineering gatekeepers. These foreign-born students seem to agree with organized American education that some of the best science and engineering graduate education available in the world is in the United States. Further, these good minds come to wealthy institutions (the research universities and the larger corporations) where they usually have ready access to the latest in scientific equipment and technology, an advantage they would not have in their home countries.

The combination is a potent one, at least in the short run. If the really able foreign-born S/Es insisted on doing their graduate work in their own countries, they probably would succeed, over the years, in raising its quality, perhaps to U.S. and Western European levels. If the U.S. were to make a major effort — including an overhaul of the current incentive structures — it could probably recruit enough first-rate native-born talent to fill our science and engineering graduate schools. But until those days come, the status quo is probably quite good for science and engineering.

---

[148] The author is grateful to Dr. Morris for calling this debt-level data source to his attention.

Chapter Six

U.S. Reactions to the Impacts of Foreign-Born Scientists and Engineers

There are two sets of reactions to the impacts of foreign-born S/Es described in Chapter Five. There is a mainstream reaction of acceptance on the part of the American Establishment and an institutional apathy on the part of those entities speaking for women, Blacks, and resident science and engineering professionals who are sometimes adversely affected by the presence of large numbers of foreign-born S/Es. There have also been some recent eddies, that is, other and stronger reactions to these impacts, that are larger than ripples but smaller than waves.

The Mainstream Reactions

Supporting the Status Quo

The large-scale presence of foreign-born S/Es, particularly at the PhD level, was neither deliberately created by America's universities and corporations nor thrust upon them against their will. For the most part it just happened. It started slowly and grew, and the establishment found it good.

Since criticisms of the phenomenon were few and weak, universities and corporations simply accepted the status quo without too much thought;[149] there was no need to mount a spirited defense or to create an implementation strategy. Externally there was a need to prevent complications arising from Congress and INS, and internally in academe there was a need to smooth out the few rough spots, such as providing language training to foreign-born TAs.[150]

When corporations and universities do rise to defend the practice, they typically use liberal rhetoric to support a conservative position. They cite "freedom of inquiry," "world-wide flows of information," and the utility of "joining the world's best minds with the world's best education." They do *not* mention the fact that the presence of the foreign-born S/Es makes life easier and more prosperous (as outlined earlier) for America's managers.

There has also been widespread establishment repetition of the notion, which will be discussed subsequently, that there was a persisting shortage of scientists and engineers, and hence a growing supply of foreign-born S/Es was needed to meet a shortfall.

At some levels, however, the scientific and engineering establishment has paid close if not critical attention to the subject and has collected many statistics on the foreign born and science and engineering. Further, it has funded studies and held conferences, which usually involved PhDs in the social sciences poking around, ever so gently, into the graduate education of PhDs in the hard sciences and engineering.

But as Jules LaPidus, president of the Council of Graduate Schools, told me in an interview on the impact of foreign-born graduate students on science and engineering graduate education for women and resident U.S. minorities, "This is not an educational problem; it is not that we are not teaching good science or that the international students are complaining. If anything it is a social or a

---

[149] Goodwin and Nacht, from another point of view, note the lack of conscious policy making regarding the presence of foreign students on American campuses. See the citation at footnote 46.

[150] A negative, but telling, indication of this lack of concern at the highest reaches of academe was the decision by Bowen and Rudenstine not to deal with the subject in their otherwise comprehensive and thoughtful study of graduate education, the previously cited *In Pursuit of the PhD*.

political problem."[151]

LaPidus is right; it is generally the noneducational aspects of the presence of the foreign born in academe, and later in the corporate world, that are the cause for concern.

Muted Opposition

Meanwhile, for at least three reasons, the presence of the foreign born in science and engineering has not created much of a political problem for the establishment. These seem to be the factors:

1. Securing more science and engineering PhDs is low on the list of priorities for organizations representing Blacks, Hispanics, and women;

2. Scientists and engineers, as individuals and in groups, are not very skilled at public advocacy and do particularly badly when representing their own economic interests;

3. An unwillingness to discuss possible linkages between immigration, on the one hand, and diminishing opportunities in science and engineering for residents, on the other, pervades many of the scientific and minority rights organizations.

In the following paragraphs, we discuss these three factors in more detail.

Not a Priority Issue. The indirect competition between foreign-born males seeking advanced education in science and engineering, on the one hand, and women and minority group members on the other, is not a hot-button issue for the latter. The numbers of people potentially involved are relatively small, and the alternative fates for those who do not pursue a science or engineering PhD are not dreadful. Further, the problem is complicated, and the rewards come years to decades later. Meanwhile, the lack of jobs generally, and reproductive and civil rights issues are all much more pressing.

---

[151] Interview with Jules LaPidus, May 5, 1993.

Lack of Advocacy Skills. Most lawyers and physicians, without losing their status as professionals, manage to do well for themselves financially; further, their organizations, such as the American Medical Association and the American Bar Association, are highly capable of defending their members' self-interests. Farmers' lobbies and plumbers' unions effectively make economic demands for their members, but organizations of scientists and engineers are virtually useless in this regard. Why is this true?

First, many S/Es lack the social skills needed for group advocacy; most scientists and engineers spend much more time interacting with concepts, numbers, and equipment than with other human beings. As one engineer put it:

> "A major...deficit that some technical workers face is a collective lack of political and social savvy. I challenge anyone to compare the legal and social savvy of professional engineering organizations to that of the AMA or the Bar Association..."[152]

Second, there is a mismatch between the self-perception of S/Es and their location in the work place. They see themselves as professionals (and most feel that professionals do not join unions), yet 99% of them are employees of large organizations. Most lawyers and physicians do not have this problem; they are either self-employed or work in partnerships and find it easy to join guild-type organizations like the AMA.

Third, the internal structure of most of the science and engineering organizations seriously inhibits effective economic advocacy, or immigration advocacy, for its members. Typically the science or engineering professional organization represents a specific technical field, such as mechanical or civil engineering, and these technical organizations tend to focus on technical subjects. While many physicians and lawyers belong to specialty organizations, most also belong to the umbrella organizations, the AMA and the ABA. There is no comparable umbrella group in either science or engineering.

---

[152] A letter to the editor, signed by Henry Bergner, who identified himself as a stress engineer from Redmond, Washington, in *Contract Employment Weekly*, Nov. 3, 1993, Kirkland, Washington, p. A-10. This is the trade paper of the contract engineers, people who hold short-term engineering jobs.

The technical organizations, particularly in engineering, do not represent just the working engineers; they also have built-in representation from engineering management and from the engineering schools. This tends to mute the organizations' ability to represent the bench engineers' interests vis-a-vis those of their bosses.

Finally, many of the organizations are international in nature, which tends to discourage them from taking stands on behalf of resident U.S. engineers, for example, that would be detrimental to either overseas or immigrant engineers.

The exceptions to these generalizations about science and engineering organizations are few. There is a relatively small organization of working engineers, based in Fort Worth, the American Engineering Association; most of its members are contract engineers or, as they call themselves "shoppers," whose careers consist of a series of short-term tours of duty with different organizations. Its president, Bill Reed, is one of the few such leaders to testify on immigration issues; AEA was opposed to much of the 1990 Immigration Act. There are also a few units of the AFL-CIO representing engineers, such as some of those employed by NASA, and there is the Professional Employees Department of the national AFL-CIO.

Organizations' Unwillingness to Tackle Immigration Policy. The final reason for the weak stance of those adversely affected by the presence of foreign-born S/Es is the unwillingness of many of the affected organizations to speak up for their own membership when it puts them in conflict with a welcoming immigration policy. This is an attitude that one sees more in the scientific than in the engineering organizations, more with those of the Hispanics than those of the Blacks.

An example was provided by a conversation with a ranking person at one of the mathematical associations. The organization's most recent employment data had shown that 13% of a recent group of PhD graduates were still unemployed almost a year after the receipt of the degree and that a substantial majority of the best jobs (at PhD-granting institutions) had been filled by non-USC applicants. In the light of these two statistics, I asked about the organization's immigration policy. The reply was, "I will talk with you about unemployment. I will talk with you about immigration. But not in the same sentence."

Thus we have an imbalance of forces: both momentum and status quo on the side of the large, well-established entities that benefit from the presence of the S/Es, while on the other side a series of often nonsubstantive factors severely limits the efforts for those disadvantaged by the presence of the foreign-born S/Es.

Universities Outmaneuver Governments

That, however, is not the end of the story. There is also the remarkable resilience and creativity of the universities when facing a government decision seeking to regulate their relations with foreign-born graduate students. Time after time, academe deflects, softens, or outmaneuvers government policies on these matters; it is a reaction to a reaction to an impact, if you will.

There was, for example, the previously mentioned "Golden Arches" Amendment to the Immigration Act of 1990, which gave INS the impression that Congress wanted F-1 students to work in the summer in only one of two ways: on campus or under the McDonald's-type program. The universities' position, which was probably correct, was that Congress was content with the existing policy on summer jobs for foreign students and that the McDonald's provision was simply yet another avenue for summer jobs. While this was being sorted out in the late spring of 1992, many universities realized that there would be very few off-campus job opportunities for F-1 students that summer, and knowing that citizen and PRA students could work anywhere, reportedly decided to save the on-campus jobs for the F-1s.[153]

Similarly, a few years earlier Congress decided that direct fellowships for graduate study would be limited to USCs and PRAs. While this decision may have been hurtful to some foreign-born graduate students in other disciplines (and effective from Congress' point of view), it made little difference in science and engineering. Universities simply used research grant funds, which contain no such language, to fund the foreign born and encouraged USCs and PRAs

---

[153] For more on the controversy over the regulations, see "INS Publishes F-1 Student Rule Reinstating Work Programs," *Interpreter Releases*, July 27, 1992, p. 898 and p. 911 The decision to move F-1 students into the on-campus jobs was much discussed at the previously mentioned 1992 NAFSA convention.

to apply for the fellowships. An effort by the late Congressman Paul Henry (R-Michigan) to give U.S. residents preference for RA appointments came to naught.[154]

Meanwhile, in Canada, various government agencies have decided, apparently on a province-by-province basis, to increase the tuition for non-Canadian students by applying an additional fee for foreign students.

In Nova Scotia, the extra fee comes to $1,700 (Canadian, about $1,225 US) per year, on top of a basic tuition of $2,800 (Canadian, about $2,016 US) per year. But for the foreign student without medical insurance there is a hidden trade-off. After paying for his or her own medical insurance for one semester ($525 Canadian), the student becomes a resident of Canada, and as long as the student does not leave Canada for more than 30 days the student receives free medical coverage.[155]

So though an American parent might sense that the additional fee is not very burdensome, many universities have found ways around it. For example, Dalhousie, in Halifax, N.S., routinely absorbs $500 of the $1,700 differential at the graduate level. Several departments, notably in science and engineering, absorb up to an additional $800 per year, leaving as little as $400 as the effective differential.

The Board of Regents of the University of California has made a similar effort to secure additional tuition from foreign students; they voted a $3,500 tuition for in-state students and a $11,050 one for out-of-state students. A U.S. citizen, after a year in California, can convert from out-of-state to in-state, but a non-USC cannot. As a result, many UC campuses have adjusted their assistance packages to foreign-born students to help them bridge this gap.

---

[154] The bill, HR 4595 in the 102nd Congress, was the subject of a hearing before the Science Subcommittee of the House Committee on Science, Space and Technology on April 24, 1992. It did not advance further. It would have required universities holding federal research grants to give preference to USCs and PRAs when making appointments of research assistants. For the bill's rationale, see the Congressman's statement in the March 27, 1992, *Congressional Record* (p. E866), and for a critique of it, see NAFSA's testimony at the hearing.

[155] The author has first-hand knowledge of this because his youngest son decided to attend a Canadian university; even with the additional fee for foreigners, higher education is so thoroughly supported by government that the costs are, from a parental point of view, wonderfully below those in the United States.

Simply passing a piece of legislation designed to reduce subsidies to foreign students, in short, does not necessarily bring about the result desired by legislators.

Eddies

The placid scene outlined above was disturbed three times in 1992 and 1993 by nonmainstream events: a congressional hearing, the forced withdrawal of one Labor Department proposal, and the presentation of another one.

Congressman Wolpe Holds a Hearing

The one-day, one-issue congressional hearing that achieves its goal is a rare and wonderful event, a jewel among the dross of Capital Hill activity.

Such a hearing was conducted on the subject of the National Science Foundation's report predicting a shortage of scientists and engineers. The hearing was before the Subcommittee on Investigations and Oversight of the House Committee on Science, Space and Technology and took place on April 8, 1992. Then-Congressman Howard Wolpe (D-Michigan)[156] chaired the meeting, which was also attended by the ranking minority member, Sherwood L. Boehlert (R-NY). Edith Holleman headed the subcommittee staff.

NSF's position was that U.S. demographic trends (i.e., a falling number of people of college age) would lead to a shortage of scientists and engineers. NSF contrasted this shortfall with what it said would be a steadily growing demand for science and engineering talent.

The NSF posture was totally destroyed in the course of the day, with the headline on the front page of the next day's *Washington Post* saying it all: "Scientist Shortfall a Myth; NSF Study Seriously Flawed, Panel is Told."[157]

Wolpe and Boehlert, who appeared to have the same point of view on this subject, listened as a long list of witnesses from academe

---

[156] He subsequently left the House because of redistricting.

[157] See *Washington Post*, April 9, 1992, p. 1.

and from the technical societies attacked the methodology of the NSF study. The study was faulted for a number of reasons, such as unacknowledged recent reductions in science and engineering employment because of corporate downsizing and defense cutbacks, the visible rise in science and engineering unemployment, the lack of recognition of the large and growing foreign-born science and engineering labor force, and the like.

Further, career staff from the Bureau of Labor Statistics discussed the problems with projecting labor demands in the future, and those from the General Accounting Office talked about how governmental studies are usually subject to peer review and to formal publication. The NSF "study" had not been subject to peer review and had never been formally released by the agency, but various photocopied versions of it had been leaked to the press and interested parties over a period of years.

Several witnesses suggested that the hidden rationale behind the report was an effort to secure additional funding for NSF and for Big Science generally. It was also pointed out that the NSF report had been frequently cited during the debate on the 1990 Immigration Act, to support more visas for skilled workers. [158]

While there was relatively little discussion of foreign-born S/Es during the course of Wolpe's hearing, a Republican state legislator from Michigan made the point that the study seemed to be built on the premise that immigration did not exist in this field. The legislator, Stephen Dresch, a demographer and a professor at Michigan Technological University, suggested NSF's priorities were misplaced, that funds should be spent on providing better education to the least well-educated half of the work force, rather than adding to the stock of unemployed PhDs.

The study's author, Peter House of the NSF Office of Planning and Assessment, and his boss, then-NSF Director Walter E. Massey, were given a tough cross-examination by both members of Congress.

And, to the best of our knowledge, the once oft-quoted study has

---

[158] See the testimony of, among others, Dr. Kevin Aylesworth of the Young Scientists' Network in: *Projected Science and Engineering Personnel Requirements for the 1990s: How Good are the Numbers?* Hearing before the Subcommittee on Investigations and Oversight of the Committee on Science, Space and Technology, U.S. House of Representatives, 102nd Congress, Second Session, No. 173, April 8, 1992.

not been mentioned in print since.

The Department Withdraws a Proposal

A much more narrowly-focused reaction to the prospect of more foreign-born S/Es in the U.S. work force occurred in the spring of 1993 when the U.S. Department of Labor proposed to "streamline" the labor certification process for employers seeking foreign workers in ten occupations, six of which were in science and engineering.

As background, the staff of the Department had long felt that there must be an easier, less-expensive way to handle labor certification requests; in many cases it was clear that there were far too many Americans capable of handling a given job to consider granting a certification, and, similarly, in many cases there appeared to be no American candidates at all for the occupation in question. Instead of handling individual cases at these two opposite ends of the spectrum, why not make use of labor market information (LMI) and rule out all applications within a given set of occupations (generally low skill ones) and accept all applications in another set of occupations, generally more skilled ones. Meanwhile, for the majority of jobs, in the middle of the spectrum, the existing individual labor certification procedure would continue.

Given the high rates of labor certification approvals (for the higher level jobs), the proposal appeared to be attractive on its face. The House immigration subcommittee, under the brief leadership of Congressman Bruce Morrison (D-Conn.)[159] incorporated the LMI notion into the 1990 Immigration Act, but only on a pilot basis.

Time passed. The Department of Labor contracted with Professor Malcolm Cohen of the University of Michigan to check out various labor market information systems to see if the Department could conduct such a pilot program. Cohen created a suggested program, DOL liked it, and in a period of interregnum (after the Bush political appointees had left but before the Clinton people had really arrived), the Department published the proposed regulations

---

[159] Morrison left the House in order to make what turned out to be an unsuccessful race for governor of Connecticut in the fall of 1990.

for the LMI-short-circuit of individual labor certifications.[160]

The LMI proposal dealt with only one of the variables suggested by Congress, shortages, excluding the concept of surpluses. The Department proposed what it regarded as ten shortage occupations, noted the job requirements for each, and listed the states in which the shortages were said to exist. The proposed system would allow employers in the states in question, seeking workers for one or more of the specified jobs, to receive an automatic labor certification, without any need to make an effort to seek a resident alien or USC worker for the job. The employer would then take the automatic certification to INS, specify the alien worker desired, and that worker would be granted an immigrant visa. The ten occupations and their job requirements are shown in Exhibit Sixteen.

Then, by the normally quiescent standards of the science and engineering labor markets, all hell broke loose. Organizations that in their entire history had *never* taken a stand on an immigration matter (e.g., the American Institute of Chemical Engineers) wrote angry comments to the Department. Trade papers, normally promanagement, ran headlines like:

CHEMICAL JOB SURPLUS ALLEGED
Finding riles scientists and Congress[161]

Individual engineers and many of their organizations wrote to the Department and to their representatives in Congress, who wrote to the Department as well. Negative stories began to appear in the general media.

Why all the excitement? There were not that many jobs at stake. How many openings were there for PhD mechanical engineers in Arizona and Texas (one of the categories proposed by DOL)? There were several reasons for the uproar:

- Institutional klutziness: none of the science and engineering organizations had been warned, much less consulted, by the Department prior to the announcement; similarly, Cohen had not

---

[160] *Federal Register*, Vol. 58, No. 52, March 19, 1993, pp. 15242-15250.

[161] *Chemical and Engineering News*, April 26, 1993, p. 6.

Exhibit Sixteen: Occupations Covered by The Department of Labor's Proposed LMI Program, An Easing of the Labor Certification Program

Category	States Covered	Degree Requirements
Biology	17	PhD
Chemistry	6	PhD
Chemical Engineering	3	MS + 2 yrs or PhD
Computer Science	7	MS + 2 yrs or PhD
Cook, Specialty	8	n/a
Materials Engineering	4	MS
Mechanical Engineering	2	MS + 2 yrs or PhD
Medical Technology	4	BS
Physician	W. Va. only	MD + licenses
Teacher, Spec. Ed.	N.Y. only	BA or BS + certificates

Note: Jobs in these occupations, with these requirements, in these states, could be filled with alien workers by employers without filing the customary individual labor certification requests should the proposed regulation be adopted by the U.S. Department of Labor.

Source: *Federal Register*, March 19, 1993, pp. 15248-15250.

---

used any of the profession-specific data sources in his work, using instead broader indices;

- Timing: between the periods covered by Cohen's research and the moment of announcement, much time had passed and science and engineering unemployment rates (while modest by factory-worker standards) had soared by professional standards;

- Symbolism: while corporate downsizing, the loss of defense-related jobs, and for some a sense of the growth in competition from foreign-born colleagues had all been nibbling away at job opportunities, there had been no single, easy-to-understand target for the frustration felt by many scientists and engineers (and their societies). The LMI proposal turned out to be the lightning rod, and as a result it attracted much negative attention.

Then, too, there was the substance of the issue. There may not have been many openings for PhD mechanical engineers in Texas and Arizona, but when a faculty position in science or engineering, anywhere in the country, is widely advertised, the sense in the professions is that there are hundreds of applications, many submitted by highly qualified candidates.

Further, the Department's proposal seemed to suggest, with its state-by-state tests of labor markets, that the labor markets in science and engineering are geographically defined, as they may well be for blue collar occupations. The science and engineering organizations felt that such jobs were routinely filled by the workings of a *national* labor market.[162]

Labor Secretary Robert Reich, who may not have had much knowledge of the proposal before it hit the *Federal Register* pulled away from the proposal, without killing it outright. Acknowledging the drum roll of protest, he asked the Congress to change a single word in the 1990 Immigration Act, substituting "may" for "shall" in that part of the law discussing the LMI pilot program and making the program optional, not mandatory. Congress subsequently repealed the entire LMI provision.[163]

Nevertheless, the point was made, in this eddy of protest, that there are those who do not want to make it even easier to bring in foreign-born S/E talent.

---

[162] See, for example, the June 1, 1993, submission of the American Association of Engineering Societies (a coalition organization) to Carolyn Golding, Acting Assistant Secretary of DOL's Employment and Training Administration, which said in part: "Adding engineers to a "shortage" list developed through a statistical analysis based on local and state data ignores the fact that engineering employment must be studied within the context of a *national* market." (Emphasis in the original.)

[163] Section 281 of Immigration and Nationality Technical Corrections Act of 1994; see *Interpreter Releases*, Oct. 17, 1994, pp. 1380 and 1400.

The Department Makes a Proposal

Late in 1993, the Department of Labor announced that it had taken a series of actions to enforce the labor market aspects of the H-1B program; at the same time it proposed a somewhat tougher set of regulations for this program.[164]

Although the Department had started some of its investigations earlier in the year, the proposed regulations (probably co-incidentally) appeared in the *Federal Register*[165] just four days after the CBS-TV expose on "60 Minutes."

According to *Interpreter Releases*, DOL had cited three computer-related firms (as well as some in the physical therapy business) for misusing the H-1B program. The largest DOL target was the Fortune-500 firm Digital Equipment Corporation (DEC), on which it assessed a $37,000 fine plus $85,035 in back wages for some 42 programmers. DEC was charged with inappropriate LCAs, which showed a salary range of $30,000 - $40,000 a year to work at various DEC sites in Massachusetts.

The most notorious of the DOL targets was Complete Business Solutions, Inc., a firm that had figured in previously cited newspaper accounts. Amerisoft, Inc., was also charged with H-1B violations, was fined $16,750, and was told to pay $23,591 in back wages to five of its foreign-born workers. While it is better that DOL be enforcing the law instead of ignoring it, the fines and back wages are small change, and the only press coverage of these decisions we have seen appeared in *Interpreter Releases*.

The proposed DOL regulations appear to be sensible, modest steps forward to remove some of the most obvious difficulties with the current program. For example:

- H-1B employers (some of whom were not paying the wages they claimed to pay in the LCAs) would have to provide a copy of the LCA to the foreign-born worker.

---

[164] "Labor Dept. Cracking Down on Alleged H-1B LCA Violators," *Interpreter Releases*, October 8, 1993, pp. 1325-1331.

[165] See *Federal Register*, Vol. 58, No. 192, October 6, 1993, pp. 52152-52164.

- H-1B employers (one of whom filed a single LCA for a proposed 11,000 H-1B workers, of which he actually hired 84) would have to file LCAs confined to a single occupation in a single area.

- Tighter standards would be set when employers set prevailing wage rates, and for informing foreign-born workers of how these rates were set.

- In-kind wages would not be allowed to count towards the prevailing wage.

- DOL would be able, under the proposed regulations, to investigate cases of possible violations without a complaint, which was not previously the policy.

- DOL also proposes tighter rules in the unlikely case of a strike in the H-1B-using facilities, and a tighter definition of the role of job shops.

In the following chapter we suggest some more sweeping changes in the H-1B program.

Perhaps the Wolpe hearing, the LMI scrap, and DOL's minor progress with the H-1B program are mere eddies in the smooth flow of mainstream reaction to the soothing presence of large numbers of foreign-born S/Es. Or perhaps they are indications that the mainstream will be challenged much more forcefully in the future.

Chapter Seven

Conclusions and Recommendations

Conclusions

We have shown that foreign-born S/Es who arrive one at a time (the immigrant pattern) generally do well in the United States. They apparently are at least as well qualified as their native-born peers when they enter American educational institutions; they secure high quality educations, mostly at the graduate level; and in later years, when they work in the U.S., they earn a bit more than their native-born peers.

Their contributions to science and engineering and to the American economy are widely recognized. Further, their presence and growing numbers are soothing to the American educational and business establishments, permitting the status quo to continue without the awkward adjustments that would be needed were they not here.

Meanwhile, there are some drawbacks to their presence. On the educational front, their availability and attractiveness to graduate schools have dampened efforts to recruit more Americans, generally, into science and engineering and more specifically, to reach out to nontraditional sources of USC S/Es, such as women and members of resident minorities. The foreign-born graduate students get more than their share of U.S. funding during graduate school — and U.S. Blacks get less than their share.

In both industry and academe, the immigrant foreign-born S/Es swell the numbers of highly qualified people for postdoc and research positions, dampening the upward adjustment in wages that probably would have occurred had there not been a surplus of such workers. Their presence has made it possible for America to avoid discussing any attempts to steer money away from financiers, industrial managers, and physicians and towards scientists and engineers.

While we see these mixed results from the immigrant pattern, the results are considerably less mixed in the nonimmigrant pattern.

Clearly foreign-born S/Es who work in the U.S. on B-1 visas (those visas designed for visiting business people) are both exploited and depressing professional wage levels where they work. Their presence is also illegal. The question is how to enforce the law in this situation.

Although the scheme is a legal one and the levels of exploitation are less pronounced, there is much to worry about in the mass movements of H-1B (those visas designed to admit people to work temporarily in the U.S.) nonimmigrants into the U.S. labor market. The question is how to minimize the abuse in this program, while still allowing corporations and universities to use this program to hire — one at a time — skilled S/Es who may be on the way to immigrant status.

Thus we have a complex situation in which a variety of interwoven governmental and private systems are producing many benefits while simultaneously creating some problems. Can an approach be devised that will preserve the useful elements and sort out the harmful ones?

An Alternative Vision

A specialized, alternative vision would look something like this:

- Incentive systems within American industry would be such that a larger percentage of Americans would be lured into productive careers in science and engineering, as opposed to equally productive careers in medicine or less productive careers in law and finance; science and engineering postdocs would, for instance, be paid at least as well as a members of the United Auto Workers;

- The incentive system would not be distorted, as it is now, by rapidly increasing numbers of highly qualified foreign-born S/Es;

- Given a stable population of foreign-born S/Es (there is no need to encourage the current cohort to leave), American industry, academe, and government would spend more managerial and economic resources to recruit Americans, of all kinds, into careers, at all levels, in S/E;

- The U.S. would continue to attract foreign-born graduate students in science and engineering, but a larger portion of the costs of their highly expensive education would be borne by those students or their home governments;

- The U.S. would then be in a situation where its own people would be sharing in a specialized form of what some might consider unattractive work, studying for PhDs in S/E, not leaving it to a foreign population;

- The U.S. would face no danger of a sudden loss of a major portion of its science and engineering intellectual elite if, for example, the economy of Mainland China were to take off (as have the economies of Taiwan, Hong Kong, and Singapore), luring home significant numbers of the U.S. scientific and engineering workforce.

To reach this envisioned state of affairs requires some sacrifices from the American establishment, which can be expected to resist with all of its strength and manipulative skills. The populations that would benefit from these reforms are, of course, simultaneously more numerous and less powerful than those benefitting from the status quo.

What needs to be done falls into four categories, described more thoroughly below:

- Adjust the system of economic incentives;
- Mildly adjust the immigrant pattern of foreign-born S/Es;
- Drastically adjust the nonimmigrant pattern; and
- Drastically adjust the pattern of self-advocacy of all resident scientists and engineers, native born and foreign born.

Recommendations

Economic Incentives

Too many Americans with good minds and solid undergraduate training are being lured into graduate programs for well-paid careers as physicians, lawyers, and financial managers; the presence of too many doctors leads to extraneous medical procedures (not reduced fees) and that of too many lawyers leads to extraneous litigation (not reduced fees). Thus, slowing down the growth in these fields would meet a real public need, quite apart from scientific and engineering considerations.

To some extent the academic establishment has moved in the right direction on this point; it is much easier to secure graduate school funding for PhDs in science and engineering than it is for MDs, LLBs (or JDs), or MBAs. But more needs to be done.

One possible answer is both remarkably simple and hard to achieve: adjust the corporate income tax code to give corporations additional credit for research and development activity while simultaneously increasing taxes on moneys secured through law suits. This would make corporate managers think more fondly of funding the R&D that this country needs so badly.

It should be noted that there are substantial precedents for this differential treatment of corporate expenditure and incomes. The cost to a corporation (or a self-employed person) of entertaining a client at lunch is treated considerably less sympathetically than the cost of renting a hotel room. Further, the IRS code was amended recently barring the use of payments to executives of more than $1,000,000 a year as an allowable business expense. Similarly, on the income side, earned income is taxed at a higher rate than capital gains.

Mild Adjustments in the Immigrant Pattern

Assuming that the major thrust for change would come through changing the incentive structure, thus attracting more native-born talent, and assuming that much of the international flow of science and engineering talent is useful to the U.S. if not to the rest of the world, only minor adjustments are needed in the immigrant pattern of individual foreign-born S/Es described in this volume.

We would recommend continuing to admit foreign-born graduate and undergraduate students as nonimmigrants, we would suggest some continuing federal funding of their graduate educations, and we would suggest only minor modifications of the labor certification program, which allows those who want to stay to adjust to immigrant status. The following should be considered:

Mandatory Two-Year Return Home in the J-1 Program. The general notion that exchange scholars should return home for two years after completing study in the U.S. should be revived; it has largely been disabled because of foreign policy considerations.

The basic idea is that the J-1 is an *exchange* program, to facilitate meaningful, short-term U.S. sojourns by talented foreigners. The objective is to give them an experience in *this* country that will help them perform in *their* country. It is designed to enhance the image of the U.S. in the minds of an overseas elite. All of these goals are, of course, crushed if the individual stays in the U.S.

These goals have been forgotten, however, in a series of piecemeal decisions. For example, since India has agreed to the waiver of the two-year homeland residence requirement, its J-1s are not required to return home; in this case, the U.S. bowed to the wishes of a friendly nation. On the other hand, because of the Chinese government's actions following the 1989 Tienanmen Square uprising, the U.S. decided that it will not force any Chinese graduate student (no matter how apolitical) to return to China. Neither of these decisions was made with any consideration of the U.S. labor market for S/Es, which is now clearly full to overflowing.

The two-year requirement should be waived only on an occupation-specific finding that, contrary to the current situation, there is a pronounced shortage of the skill concerned in the U.S. This should be a U.S. government decision, not one made by a foreign government. Care also should be taken to issue waivers on the grounds of marriage to USCs only in the case of a true hardship situation. Two years in Korea, for the USC spouse, for example, should not be regarded automatically as a hardship, as it is now.

Tighter administration of this provision will presumably be welcomed by those, here and abroad, concerned about the Brain Drain. Similarly it would meet the needs of some foreign governments some of which are not pleased by the extent to which their S/Es stay in the U.S.[166]

Funding Science and Engineering Graduate Educations. We have noted that nonimmigrant PhD candidates are somewhat more likely to be funded by U.S. sources than native-born ones and that they are much less likely to be in debt upon graduation than their native-born peers, particularly Blacks.

With this in mind, we recommend that all U.S. government research grants carry a stipulation that no more than 25% of the funds within any such grant for students and postdocs' salaries and stipends be used for noncitizens. Since noncitizens currently make up a slightly higher percentage of science and engineering grad students than 25% and a much higher percentage of postdocs, this would encourage the universities to (a) support and recruit more USC graduate students, (b) stimulate immigrant graduate students (a relatively small group) to become citizens, and (c) rethink the institutions' normal posture on *not* making loans to foreign-born grad students. We suspect that the flat percentage proposed here would be easier to enforce than the somewhat fuzzier approach of the previously described Henry bill. Currently all U.S. government graduate fellowships, a rather less significant funding channel than research grants, go to USCs and PRAs.

These proposed financial arrangements would tend to restore a level playing field, which is currently tipped, if slightly, in favor of the foreign-born graduate students.

An even better, but politically more difficult recommendation, would allow more and larger stipends to graduate students and postdocs, both native born and foreign born, and cost the Treasury less. This would be accomplished by channeling all funds currently spent on graduate student and postdoc stipends through research grants into a different funding mechanism: direct grants awarded to competing grad students and postdocs.

---

[166] Muhler notes: "The Chinese, for example, curtailed the flow of their PhD students in some scientific fields to the United States -- even prior to the May 1989 bloody upheaval in Beijing -- because a large number never returned. These students are now being sent to Western Europe... where they are less likely to find employment, permanent jobs or social acceptance than in this country [the U.S.]." Thomas Muller, *Immigrants and the American City*, New York University Press, New York, 1993, pp. 125-126.

The savings, or more and larger stipends, would come because the traditional overhead rate on pass-throughs to individual scholars is a few percentage points, as opposed to the fat overhead rates awarded for research contracts. In this way federal money would be taken away from university overhead accounts and given directly to the S/Es.

While this makes good sense, it would alienate two important populations: the managers of the universities and the current administrators of research-funding programs in agencies like the Departments of Defense and Energy. The latter would have to adjust their long-established programs to meet what they would surely regard as extraneous social objectives.

Labor Certifications. Contrary to the Labor Department's efforts with the LMI in 1993, labor certifications for foreign-born scientists and engineers should be marginally *more* difficult to obtain, as a constant reminder to the American establishment that it should be doing more to encourage better science and engineering education and better science and engineering recruitment in the United States. Such labor certification should not be impossible, just more expensive and more difficult.

We have noted the currently lopsided decision-making in this field, in which native-born workers get the decision less than 10% of the time and the foreign born win in more than 90% of the cases. The author's own background with U.S. government-operated decision-making systems regarding foreign workers suggests that if one wants to open up jobs to U.S. workers, one should discourage employers from applying for alien workers rather than tinker with the decision-making process within the U.S. government.[167]

With this in mind we would suggest the following:

---

[167] The Migrant Legal Action Program, for example, which has been fighting a lonely battle against exploiting farm workers via the use of the H-2A program (which brings Jamaicans into the vegetable and sugar cane fields) has long since turned away from trying to cause the Labor Department to make decisions on behalf of U.S. farm workers. Instead, these poverty lawyers have sued H-2A users in Presidio, Texas, in the Florida sugar industry, and in West Virginia apples, among others, under state and federal labor laws, winning most of their cases. Many growers, rather than face the resulting costs, have opted to use domestic workers.

- Labor certification fees be set high enough, particularly for multiple-usage, that employers will notice the fees and think more positively about hiring U.S. workers. A suggested annual fee schedule could be zero for the first two labor certification applications, $1,000 each for the next eight, $2,500 for the next ten, and $5,000 each thereafter. Fees for nonprofits, such as universities, could be set lower than the corporate levels.

  The fees would be established on a nationwide basis, and each corporation and university would be required to maintain its own central registry of these applications. (Again a nuisance and a small wake-up call to management.)

- Any corporation or nonprofit filing three or more labor certification applications a year would be obliged to forward copies not only to pertinent unions, as now, but to all potentially interested technical societies and trade papers. Getting these sometimes absurdly specific job requirements[168] into the hands of the trade press would probably lead to some interesting journalism.

- Professional organizations should be encouraged to pay closer attention to labor certification issues and to initiate civil suits when their members are economically disadvantaged by these actions.[169]

Clearly these proposed reforms will only slow the filing of labor certifications, not stop them, but will presumably make employers more careful about this decision. As to the extra $1,000 to $5,000 in fees to the government, this is more of an attention-getter than a financial barrier. If corporations were concerned about these costs,

---

[168] Some of these job requirements, designed to facilitate the admission of a specific foreign-born worker, are tailored so tightly that they not only require a specific list of graduate and undergraduate courses, they also specify the *sequence* in which the courses must have been taken. Conversation with Edith Holleman, staff director, Subcommittee on Employment, Housing, and Aviation, Committee on Government Operations, U.S. House of Representatives, Dec. 9, 1993.

[169] For a precedent in agriculture, see footnote 167.

they could save considerably more than these proposed fees simply by using less expensive law firms for this work or by farming it out to the international students offices of the research universities.

Drastic Changes in the Nonimmigrant Pattern: B-1

As noted earlier, the principal need in the B-1 program is to enforce the law, that is, to abolish the use of this visa to bring workers into the U.S. labor market. Given that the principal agencies concerned, the State Department and INS, have historically shown only a muted interest in the exploitation of workers in the U.S. labor market, the suggestions that follow are detailed.

State Department. The Department should devise some way to give real incentives to the young consular officers overseas who detect and refuse visas to S/Es seeking to work in the U.S. on B-1 visas.

The Foreign Service Institute should invite some knowledgeable voices from organized labor to talk about the importance of bearing in mind the interests of American workers when making B-1 visa decisions. This is a voice, to the best of our knowledge, that is not heard in the FSI training programs for consular officials. A specific discussion of the ways to detect the abuse of B-1 visas in science and engineering would be useful, too. Representatives of organized labor, similarly, should be invited to speak at INS training programs as well.

If some illicit B-1 S/Es are deported as a result of INS activity (see below), the State Department press officers in the home nation of the deportees should publicize the deportations, particularly in the relevant local trade and academic press, and, at the same time, discuss the appropriate ways to seek work in the U.S. (The point that would be made is that it is possible and legal for a computer programmer, for example, to work in the U.S. if the right kind of visa is secured; another point is that the individual alien had been badly exploited by his or her employer.)

INS. The Service should establish a small (3-person), year-long task force in the Central Office Intelligence or Investigations unit, to track patterns of B-1 abuse among programmers and to convey to the field and to consular officials how these programs work and how they can be detected. At the end of the year the task force could turn its attention to other forms of abuse of B-1 visas in the labor market.

As soon as INS finds some illicit use of B-1 visas in science and engineering, it should coordinate a broad-brush campaign against the corporate abusers, seeing to it that the trade and general press (and "60 Minutes") know about the INS actions; further, all other interested state and federal labor and tax officials should be alerted to (and lobbied for) enforcement action against the abusers, as in these cases it is certain that many laws, in addition to the immigration law, are broken. Similarly, INS should, as it rarely does, let the interested technical and labor organizations know of its work in this field. In addition, INS should publicize the deportation of the individual visa abusers both in the United States and in the abusers' homelands (see above).

Science and Engineering Workers Organizations. These organizations and those representing U.S.-based contractors hurt by the B-1 pattern should press forward with two activities unusual to them.

First, both should alert their members to the B-1 abuse and set up mechanisms for collecting detailed and useful information on these patterns (names of workers, of employers, location of consulates issuing the visas, etc.). Such information should be passed along to the State Department and the proposed INS task force, *always* with copies to members of Congress and the press to enhance the likelihood that attention will be paid to them.

Second, in instances in which it is clear that employers continue to use B-1 workers, the traditional protest mechanisms (pickets, boycotts, news conferences, etc.) should be established vis-a-vis the major U.S. economic entity that is profiting from these low-cost services.

The Congress. A suitable Congressional subcommittee[170] could, on the anniversaries of the "60 Minutes" program conduct hearings on what various government agencies had done since the program to limit the abuses of the nonimmigrant programs in science and engineering. Ideally, the agencies would be given ample notice of this hearing, so that they would be stimulated to take substantive action

---

[170] The choice is wide. In the House, for instance, such a hearing could be conducted by a subcommittee of Labor and Education, of Government Operations, of Judiciary, or, perhaps, of Foreign Affairs.

prior to it. And, ideally, for the subcommittee chair, the chair's picture would appear on "60 Minutes."

Less Drastic Change in the Nonimmigrant Pattern: H-1B

As noted earlier, the Department of Labor has proposed some useful initial steps towards reducing some of the abuses in the H-1B program. It needs to go much further, but first it must acknowledge the basic difference between the use of the program to bring in single workers (often a step in the immigrant pattern) and its use to bring in groups of workers (always part of the nonimmigrant pattern). The latter is *always* more harmful to resident workers and more harmful to the utilized nonimmigrant workers than the former. To this end, DOL and other government entities should take the following steps:

Fee structure. As with labor certification, there would be a graduated fee schedule — the more nonimmigrant workers you want, the more each costs. The fees for industry and academe for labor certification could be adopted here as well. Fees should be levied on all applications, whether or not they are approved.

Differential Scrutiny. As a matter of conscious practice, the Labor Department should announce that it will pay much more attention to applications for groups of people than single applications.[171]

If more than 10 applications are received from a single facility, for example, DOL investigators should make site visits, preferably unannounced, to talk with individual workers, both foreign born and native born (away from management's eyes and ears) about wages and working conditions. The fees to be paid by those seeking H-1B applications should be used to fund these investigations.

Notice. To make sure that all taxes are paid and to make certain that arriving H-1Bs know exactly what their actual pay checks will be employers seeking three or more H-1B workers in the course of a year should be obliged to have an exchange of correspondence with

---

[171] One potential problem, of course, is that mass users of H-1Bs will seek to camouflage their activities by creating subsidiary corporations, assigning workers to subcontractors and the like.

their would-be workers outlining wages and deductions. Currently there is no obligation on the part of an H-1B employer to tell his nonimmigrant worker about the extent of federal and state income taxes, FICA and Medicare deductions, nor the standard 30% deduction required by IRS for nonresident aliens.

Further, copies of these exchanges of correspondence should be filed, by the employer, with:

- All pertinent federal, state, and local taxing authorities;
- Unions, where present;
- Pertinent technical or professional organizations; and
- The trade press.

Prevailing Wage. The arrangements currently made for determining the minimum wage are inadequate, as is the puzzling requirement that the employer has to pay only 95% of whatever is found to be the prevailing wage. Why not 105%, to encourage the employer to pay attention to resident workers, be they citizens or legal immigrants?

Instead of using the hard-pressed and under-funded State Employment Security Agencies to establish prevailing wages, why not use the generally available, occupationally specific salary surveys of the various professional and engineering societies?

In this scenario, if the employer had wanted to hire a chemist with a bachelor's degree and two to four years' experience in 1993, for example, the employer would have been required to pay the prevailing wage for such workers as established by the American Chemical Society's annual salary survey, or $33,300. Chemists with master's degrees, or two to four years after the BS, would get $39,000. The ACS survey, as currently conducted, does not provide a prevailing wage for a new PhD, but chemists with doctorates five to nine years after the BS received $53,000 in 1993.[172]

These salary surveys reflect the workings of the national labor market and are fair to the workers involved. The salary scales are easily obtainable by employers (from the trade press) and are not vulnerable to manipulation. Neither statement can be made about the current system.

---

[172] See Michael Heylin, "Job Market for Chemists Remains Depressed, Salaries Gain 5%," *Chemical and Engineering News*, July 12, 1993., p. 10.

Why not use 105% of these prevailing wages in setting salaries for nonimmigrant workers? In that way all concerned could be assured that wage levels would *never* be depressed by the use of nonimmigrant workers. Even at 100% of the salary-survey wages employers could argue that they were not paying below the prevailing levels (even though tapping into a foreign labor force might well, ultimately, depress wages generally).

The 65,000 Limit. The 1990 Immigration Act, for the first time, established a numerical limit on the approval of a set of nonimmigrant petitions, in this case for H-1B visas. Given a piece of legislation that was a disaster, generally, for American workers competing with foreign-born workers, this was a remarkable provision. Now that there is some possibility that the limit will actually work to reduce the growth in the number of nonimmigrant workers, the immigration bar and their corporate allies are sure to seek to repeal or soften this provision.[173] These efforts should be resisted stoutly. If there is a prospect of reaching the ceiling, INS should allocate the limited number of petition approvals near the end of the fiscal year by granting them to employers offering the highest wages.

Drastic Modification in Science and Engineering Economic Advocacy

Working out an arguably appropriate public policy, however, is a small step compared to the real challenge of implementing reform in this field, where both the myth and virtually all the political power are on the side of status quo and where those hurt by the status quo are scattered, politically dysfunctional, and sometimes totally unaware of what is happening to them. Most important, the very structure of the organizations that represent the scientists and engineers are such that they are almost completely useless when it comes to economic advocacy for their working membership.

Scientists and engineers today, although better treated financially than they have been in the past, are approximately where factory workers were in the early 1930s. There were organizations within

---

[173] See, for example, "INS Approved 61,591 H-1B Petitions Last Fiscal Year," *Interpreter Releases*, November 1, 1993, pp. 1438-1439.

labor then that dealt with economic advocacy — the old craft unions — but they were unwilling or unable to organize the factory workers.

Rather than trying to modify the ways of the old American Federation of Labor, John L. Lewis (of the Mineworkers) and his younger allies (such as Philip Murray and the Reuther brothers) simply set up new organizations, under the umbrella of the Congress of Industrial Organizations (CIO).

Something like that needs to be done today. While the technical and professional societies can continue to do their useful educational work, some other entity needs to enter the picture, to lobby effectively for a greater emphasis on science and math in the K-12 schools, for better funded undergraduate and graduate education in science and engineering, and for better pay for working engineers and scientists.

There are at least three organizations that are paying attention to these issues: the Young Scientists' Network, the Professional Employees Department of the AFL-CIO, and the American Engineering Association, Inc., which speaks primarily for contract technical workers. One of the new tools available to these groups, and used extensively by the Young Scientists, is the E-mail network. Perhaps these three could, with others, establish a new umbrella organization that would speak up for the bench scientists and engineers.

Such an organization would be less likely than existing technical and professional societies to turn a blind eye to the unpleasant notion of competition between foreign-born and native-born workers. Such an organization might also tackle the even more troubling prospect, particularly in software development, that science and engineering work will simply leave the U.S. and be performed elsewhere (e.g., India) where skilled professionals abound and where wages are much lower than in the United States.

Further, such an organization might have enough imagination to sit down with the poverty lawyers who work for farm workers in order to learn from the agricultural experience how to protect the economic interests of scientists and engineers, a strangely powerless group of American workers.

But what is really needed is a towering leader, someone who can bring national attention of these issues — another John L. Lewis — but that prospect appears to be slight.

Appendix

Definitions, Data Sources and Estimating Techniques for Exhibit One

Definitions

Nonimmigrant admissions are for those classes who may work in the U.S.; in 1969 this included H-1, H-2, and H-3 (worker) visas, and J-1 (exchange visitors). By 1979 the L-1 class (employees of multinational firms) had been added. The 1993 grouping includes these classes plus those admitted under the Canadian Free Trade Agreement and a few under the new O visa (for extraordinary ability and achievement). Meanwhile, for S/Es the former H-1 class has become H-1B and the former H-2 class has become H-2B. These are the classes of nonimmigrants about whom INS collects occupational data.

The occupations used by INS since 1980 have been those of engineer, natural scientists, and computer, mathematical and operations research scientists. For 1969 the occupations were: chemists, engineers, agricultural scientists, biological scientists, geologists and geophysicists, mathematicians, physicists, and statisticians and actuaries; presumably some to many of the "professors and instructors" that year were in science and engineering but none were included in these calculations. In 1979 the occupations were: computer specialists, engineers, mathematical specialists, and scientists, life and physical; again, "teachers, college and university" were not included in the tabulations. All INS data are for the government's fiscal year, while the balance of the exhibit is for the academic year in which they end. Thus "1991" indicates graduate enrollment for the fall of 1990, and for degrees awarded throughout that 1990-1991 academic season.

Enrollment data are for nonimmigrants, while doctorate data are for noncitizens; master's and bachelor's degree data are for nonimmigrants. Generally science and engineering is defined, within these data sets, as all engineering studies and most science (excluding agriculture, psychology and social science).

Data Sources

Nonimmigrant admissions data for 1969 and 1979 are counts taken from the 1969 and 1979 *INS Annual Reports*, table 16B. In more recent years they are the author's estimates, based on INS data, calculated in a manner described below.

Immigrant data are from Table 8a of the 1969 and 1979 *INS Annual Reports*, from Table IMM.6.6 of the 1985 *INS Statistical Yearbook*, from Table 19 of the 1990 and 1991 *INS Statistical Yearbooks*, and from unpublished INS data. Data are counts.

Graduate school enrollment data (for noncitizens) are from National Science Foundation, *Selected Data on Graduate Students and Postdoctorates in Science and Engineering, Fall 1992*, table 14. All University enrollment data (i.e., graduate and undergraduate) are from Institute of International Education, *Open Doors 1990-1991*, p. 29, and *Open Doors 1992-1993*, pp. 32-33.

Doctorate awards data are from various National Research Council *Summary Reports*; 1969 data from the 1989 *Report*; 1979, 1985, and 1990 data from the 1990 *Report*; and 1991 and 1992 data from the 1991 and 1992 *Reports*. Bachelor's and master's degree data, which are *for nonimmigrants only* and thus understate the totals for the foreign-born by 20% or so, are from National Science Foundation, *Foreign Participation in U.S. Academic Science and Engineering, 1991*, pp. 76-77, for 1979-1990, and, for 1991 and 1992, from the Center for Educational Statistics by phone.

Estimation Techniques

For nonimmigrant admissions, 1985-1993: half or more of the forms filed by arriving nonimmigrants (the I-94) do not have occupational information, because the inspector has not entered it; since this is the case, we regarded the occupational distribution of those with recorded occupations as a sample of the total population and extrapolated these estimates from unpublished INS statistics on nonimmigrant admissions (table 614 in the admissions data series).

The author is grateful to Michael Hoefer, INS's Chief of Demographic Statistics, for his help with the admissions data.

Index

Adjust economic incentives 163
Advocacy 2, 135, 145, 147, 148, 163, 170, 173
Affirmative action 50, 142
AFL-CIO 174
American Bar Association 148
American Chemical Society 107, 135
American Engineering Association 149, 174
American Institute of Chemical Engineers 25, 155
American Medical Association 148
Approval percent, INS 63
Association of American Universities 49
Association of Graduate Schools 49
Authority, amenable to 12, 14, 130
Avoiding adjustments 124
Aylesworth, Kevin 74, 153

Balance-of-payments 78
Baltimore 104
Biochemistry 34, 49
Biology 34, 35, 96
Biometrics & biostatistics 34
Boehlert, Sherwood L. 152
Botany 36
Bouvier, Leon F. 6, 109
Bowen, William G. 71, 75, 125, 126
Brain Drain 1, 165

BS/BA 3, 7, 8, 10, 11, 25, 27, 31, 47, 70
Bureau of Labor Statistics 153
Business executives 3
Business Week 78

Canada 5, 38, 76, 151
Canadian Free Trade Agreement 63
Census 6, 23
Center for Immigration Studies 6
Chemical engineering 27, 34, 96
Chemical engineers 29
Chemistry 7, 34, 35
Chemists 29
Cherian, Joy 105, 109
Citizenship 102
Civil engineering 1, 34
Cohen, Malcolm 154, 156
Commerce, U.S. Dept. of 78, 79
Committee on Sci., Space & Tech., U. S. House of Rep. 152
Computer science 6, 7, 34, 42, 96
Computer scientists 117
Computer software 117
Congress 146, 150, 157, 170
Congress of Industrial Organizations 174
Council of Graduate Schools 146
Country of origin
  Africa 15

Asia 15, 40, 74
Australia 15, 117
Bangladesh 96
Canada 17, 40, 117
China 14, 15, 17, 31, 66, 87, 106, 165
Cuba 31
Europe 15, 100
Germany 17
Great Britain 17, 40, 117
Greece 17
Hong Kong 15
India 15, 17, 31, 105, 117, 120, 165, 174
Iran 17
Jamaica 2, 142
Korea 17, 165
Mexico 74
Near East 88
New Zealand 15
Nigeria 123
North America 15
Poland 31
Russia 4, 13, 118
Singapore 17
South America 15
Switzerland 31
Taiwan 15, 17
Ukraine 118

DDS 138
Demographic trends 6-8
Dentists 23
Deportations, publicized 169
Disabled 2
Displacement 128
Distribution of income 121, 125, 130, 162, 164
Downward loyalty 137
Dresch, Stephen 153

Early retirement 136
Earnings
 PhD vs. BS degree holders 27
Earth sciences 96
Economics 49
Education, total costs of 84
Educational reform 130
Educational Testing Service 40, 89
 cultural bias 39
 generally 37, 38, 45
 security 45, 46
Ehrenberg, Ronald G. 76
Elapsed Time to Degree 75
Electrical engineering 1, 96, 106
Emigration of U.S.-trained S/Es 132
Emigration rate 70
Engineering 6, 7, 11, 34, 42, 44
English language 12, 15, 35, 38, 39, 49, 74, 90, 123
Enrollment of foreign born 8, 10
Environmental science 7
Exchange program, changes in 165
Export of education from U.S. 78, 79

FICA 172
Finn, Michael 12, 18, 70, 109, 129
Foreign medical graduates 32
Foreign student advisors 66, 131

General Accounting Office 153
Geology 34
Goodwin, Crauford D. 48
Graduate schools
  admissions 3, 14, 37, 46, 47, 53, 54, 143
  attrition rate 75, 76
  cost 22, 23, 25
  debts, loans 83, 88, 91, 144, 166
  dissertation 71, 72
  entrance fees 88, 89
  financial aid 27, 30, 33, 69, 77, 86-88, 111, 141, 142, 166
  funding techniques 88
  generally 3, 8, 10, 21, 25, 27, 29, 30, 71, 73, 124
  more low-quality departments 126
  more small departments 125, 126
  qualifying examinations 71
  recruiting 143
  Who pays? 77-79, 82-84, 163
GRE 38-40, 42, 44, 45, 89, 143
Green card 10, 52, 83, 91, 104, 131
Greenberg, Daniel S. 98

Henry, Paul 151, 166
Higher wages for foreign born 106, 109, 110
Holleman, Edith 152
Hostile powers 133
House, Peter 153

Immigrant path 4, 5, 37, 101-103, 105, 106, 121, 129, 134, 161, 163, 164
Immigrants 8, 10, 18, 19, 44
Immigration Act of 1990 56, 115, 149, 150, 153, 154, 157, 173
Immigration system 3
Impact on graduate education 124, 125
Impacts on U.S. education 122
Income differentials 23
Income, distribution of 3
INS 8, 18, 31, 37, 47, 50, 52, 53, 55, 57-63, 66, 114, 115, 146, 155, 169, 170, 173
Institute of International Education 78-83
Interpreter Releases 66

JDs 3, 21, 22, 25, 32, 33, 138, 164
Justice, U.S. Dept. of 55

K-12 Education 122
Kuh, Charlotte 40, 42

Labor certification 5, 52, 56, 64, 65, 132, 154, 155, 165, 167, 168, 171
Labor certification fees, increase in 168
Labor Condition Application 56, 61, 92, 115, 118, 120, 158
Labor market information proposal 154, 155, 157, 159, 167
Labor markets (academic)

generally 2
hiring 54, 98-100, 102, 131
wages 87, 92-94, 100, 121, 123, 127, 128, 162
Labor markets (corporate)
data gaps 118
discrimination 54, 110
generally 133, 162
hiring 50, 52-54, 100-102, 131, 135, 136
wages 2, 23, 25, 106, 107, 109, 111-115, 117, 118, 129, 130, 134, 162, 171, 172
Labor markets (governmental)
"Glass ceiling" 105
hiring 103, 103, 131, 132
wages 104, 109
Labor, U.S. Dept. of 55, 65, 92, 93, 115, 118, 152, 154, 158, 171
LaPidus, Jules 101, 146
Law School Aptitude Test (LSAT) 39
Lawrence, Lauren 66
Lawyers 3, 23, 52, 131, 148, 164
Lewis, John L. 174
Life sciences 7, 11, 21, 34, 35, 42, 44, 93, 102
Lingua Franca 87

Maryland 104
Massey, Walter E. 153
Mathematicians 117
Mathematics 1, 6, 7, 11, 15, 34, 35, 42, 49, 96
MBAs 3, 21, 22, 25, 32, 33, 164

MDs 3, 21, 22, 25, 31, 33, 138, 164
Mechanical engineering 1, 34, 49
Mining engineering 96
Morris, Frank L., Sr. 39, 141
Morrison, Bruce 154
MS/MA 4, 8, 10, 11, 13, 31, 44, 70, 76
Murray, Philip 174

Nacht, Michael 48
National Association for Foreign Student Affairs 52, 66
National Association of Computer Consultant Businesses 116
National Institutes of Health 93, 94
National Science Foundation 12, 31, 48, 91, 93, 152
Natural sciences 6
Naturalization 4, 5, 12, 18, 19, 55, 57, 62, 64
Naturalized citizens 70, 98, 100, 101, 105, 106, 108
Navy, U.S.
Office of Research 72
Negative impacts of foreign-born S/Es 134
New York Times 78
Nonimmigrant path 4, 5, 101, 102, 111-113, 116, 118, 122, 129, 132, 134, 163, 169-171
Nonimmigrant visas
A (foreign diplomats) 112
B (visitors for business) 111-114, 116, 117, 162,

169, 170
F (foreign students) 57-60, 66, 150
generally 5, 57, 63, 64, 83, 90, 91
H (temporary workers) 10, 56-58, 61-64, 92, 111, 112, 114-117, 120, 162, 171, 172
J (exchange visitors) 10, 57, 58, 60-62, 165
L (multi-national employees) 10, 56-58, 62-64
Nonimmigrants 4, 8, 10, 12, 13, 18, 42, 44, 52, 55, 57, 58, 65, 70, 83, 100, 102, 105, 107, 142, 162

Overhead rates 121, 167

Papademetriou, Demetrios G. 111
Parlin, Bradley W. 54
Passport, U.S. 55
Percent foreign born 1, 5, 6, 8, 10, 11, 18, 44, 48, 70, 76, 96, 98, 126
Permanent resident aliens 4, 12, 30, 42, 44, 56, 70, 100, 102, 105, 107, 142, 150, 166
Petroleum engineering 96
Physical sciences 11, 44
Physicians 3, 23, 32, 122, 139, 148, 164
Physics 7, 34, 35, 94, 96
Positive impacts of foreign-born S/Es 133
Postdoc 62, 67, 87, 91-94, 96, 98, 100, 105, 127, 128, 166
Preference for U.S. citizens 47

Reich, Robert 157
Research and development 121, 164
Research assistants 15, 86, 88, 90, 127, 128, 142
Research grants 124, 150
Resident Minorities
  Blacks 2, 4, 11, 34, 42, 49, 50, 84, 106, 108, 121, 134, 137, 139-145, 147, 149, 166
  generally 130, 140, 147
  Hispanics 2, 4, 11, 34, 42, 49, 50, 106, 134, 137, 139, 140, 143, 147, 149
  Native Americans 134
  recruiting efforts 141, 142
Return migration 69
Reuther brothers 174
Rudenstine, Neil L. 71, 75, 125, 126

60 Minutes 102, 114, 117, 120, 158, 170
S/Es with foreign degrees 11, 12
S/Es, degrees awarded 5-8, 25
Schneider, James 116
Scientific skills, need for 139
Shortage of S/Es 122, 146
Simcox, David 6
Smullin, Louis 47
Soothing the establishment 121, 129, 146
Stafford Loans 91

State government 103, 104
State, U.S. Dept. of 55, 65, 113, 169, 170
Summary Report 1990 11

Tax code, changes in 164
Taxes 171
Teaching assistants 15, 39, 86, 88, 90, 123, 137, 138, 146
Thao, Paoze 65
Third World 1, 12-14, 89, 130
Tienanmen Square 58
TOEFL 38, 39, 45, 89
Total costs of education 84
Trade press 170, 172

U.S. school system
  chemistry 35
  physics 35
Undergraduate admissions 46
Undergraduate education 123
Unemployment 135, 136, 156
Unions 148, 149, 169, 170, 172, 174
United States Information Agency 58
Universities
  Berkeley 128
  Cambridge 13
  Cornell 90
  Dalhousie (N.S.) 36, 72, 151
  George Washington 46
  Harvard 46
  Harvard Medical 128
  Michigan Technological University 153
  MIT 13, 47, 128
  Morgan State 39, 141
  Northwestern 87
  Princeton 46
  Stanford 127
  University of California 151
  University of Delaware 89
  University of Michigan 154
  University of Nebraska 74, 93
Universities outmaneuver governments 150
University overhead accounts 124, 126-128
Unwillingness to discuss immigration 147, 149

Wang, An 110
Washington Post 78, 118, 152
Willison, J.H.M. 36
Wives of foreign-born S/Es 139
Wolpe, Howard 152, 159
Women 2, 4, 11, 12, 34-36, 106, 108, 121, 130, 134, 137-139, 145, 147
Work ethic 13, 74, 110, 129
Working with one's hands 74

Yale-Loehr, Stephen 66

Young Scientists' Network 174

Zinberg, Dorothy S. 101
Zoology 34, 36
Zydney, Andrew 89